Furneaux Jordan

The Treatment of Surgical Inflammations

By a New Method Which Greatly Shortens Their Duration

Furneaux Jordan

The Treatment of Surgical Inflammations
By a New Method Which Greatly Shortens Their Duration

ISBN/EAN: 9783337814298

Printed in Europe, USA, Canada, Australia, Japan

Cover: Foto ©berggeist007 / pixelio.de

More available books at **www.hansebooks.com**

THE TREATMENT

OF

SURGICAL INFLAMMATIONS

BY A NEW METHOD,

WHICH GREATLY SHORTENS THEIR DURATION.

BY

FURNEAUX JORDAN, F.R.C.S. Eng.

SURGEON TO THE QUEEN'S HOSPITAL;
PROFESSOR OF SURGERY AT THE QUEEN'S COLLEGE;
PROFESSOR OF ANATOMY TO THE BIRMINGHAM ROYAL SOCIETY OF
ARTISTS.

LONDON:
JOHN CHURCHILL AND SONS, NEW BURLINGTON STREET.
1870.

CORNS AND BARTLEET, PRINTERS, UNION STREET, BIRMINGHAM.

PREFACE.

In the following pages I assume (and the results recorded hereafter justify the assumption) that the inflammatory process is one and the same process wherever it may occur, and that the best remedies for any inflammation are the best in all. That drugs have, for certain purposes, great utility I do not deny. Putting aside, however, specific inflammations I know no drug, or combination of drugs, that can cure an inflammation. Counter-irritation can unquestionably cure inflammations. Let me cite one fact among many. An inflamed patellar bursa, before chronic thickening sets in, can be most certainly cured by a ring of blister placed around it, when every other remedy

fails. Seeing that counter-irritation could do
what no other remedy could, I put a ring of
counter-irritation around abscesses, carbuncles, cellulitis, and other inflammations, and
discovered striking and rapid results. I next
sought to find out the most favourable
localities for counter-irritation in *every* inflammation.

It appeared to me that it should always be
established over the next, or another, vessel
or vascular territory.

I knew that even and gentle pressure could
rapidly arrest some, especially chronic, inflammations. For removing the products of
inflammation I knew no remedy to be compared with it. I used it, therefore, in all
accessible inflammations.

The best mode of using the best remedies
(counter-irritation, pressure, rest, and a few
others) in every inflammation is not a routine
treatment. Take counter-irritation alone;
there is endless scope for individual and in-

dependent judgment in the determination of its best localities, best degrees of intensity, best extent and configuration of area, and best modes of production. It is the same with the other remedies advocated in these pages.

One of the greatest advantages of counter-irritation is this—it immediately and efficiently relieves inflammatory pain. Most patients are eager to exchange a continuous, dull, wearing, inflammatory pain for a temporary smarting.

The cases are very briefly reported, as they are intended simply to illustrate the effects of treatment.

I beg here to thank numerous medical friends for the interest and care they have shewn in making trial of the treatment. Many gentlemen adopt it in all inflammations, others adopt it in some, a few believe it to be the best treatment for orchitis, others that it is the best for enlarged glands, others that it is the best for carbuncle, and so on. I am indebted to Mr. Turton, of Wolverhampton, Mr. Hickinbotham, Dr. Hodges, Mr. S. Lloyd, and Dr. Quirke, for several interesting cases. Dr. Earle informs me that he finds the treatment here described of very great benefit at the Children's Hospital. My colleague, Mr. Wilders, of the Queen's Hospital, and Mr. Newnham, surgeon to the Wolverhampton Hospital, both assure me that they frequently adopt the treatment, and find it singularly rapid and efficient. I have heard through friends that many surgeons in different parts of the kingdom resort to it. From these gentlemen I should be glad to receive abstracts of cases, opinions, and suggestions.

I am especially indebted to my friend Dr. Sawyer for valuable service in the revision of the following pages, and for many useful suggestions.

COLMORE ROW, BIRMINGHAM.
OCTOBER, 1870.

CONTENTS.

PLATES.

The following diagrams are mostly drawn from actual cases. The dotted lines show the extent of counter-irritation, of a moderate character, as effected by iodine pigment. Unless otherwise indicated, the whole circumference of the limb is irritated.

PLATE I.

Fig. 1.—Extent of counter-irritation in a large abscess of the fore-arm. If the first application of iodine is efficient, the relief of pain is striking and immediate. See Mr. Hodges's case of Mammary Abscess, and Mr. Turton's cases of Carbuncle.

Fig. 2.—Shows singular appearance, after twenty-four hours of counter-irritation, in removing inflammation, and circumscribing pus. I have several times seen this tumour-like collection of pus removed in a few days by continued counter-irritation.

Fig. 3.—Extent of counter-irritation (iodine) in abscess of popliteal space. An iodine paint is applied to the whole circumference of the limb within the dotted lines.

Plate 1

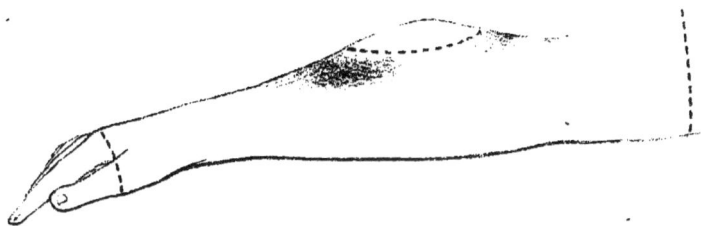

Fig 1

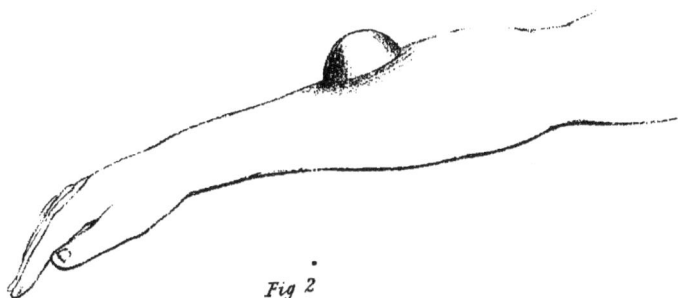

Fig 2

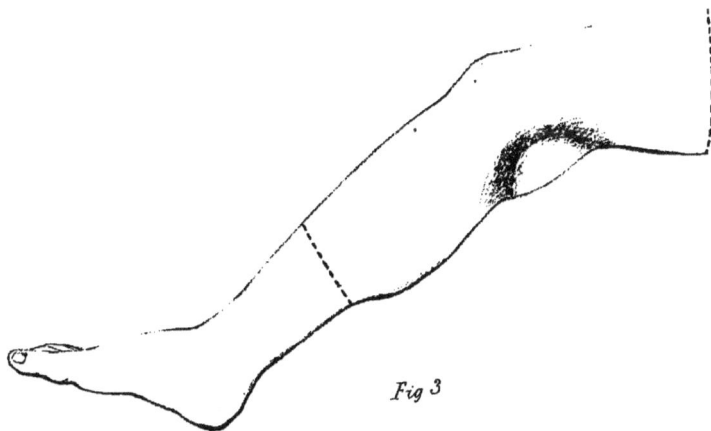

Fig 3

PLATE II.

FIG. 1.—Extent of counter-irritation in abscess of neck. The counter-irritation should be active (iodine liniment occasionally, or paint frequently) in so limited an area.

FIG. 2.—Counter irritation in abscess of axilla. The dotted line is supposed to be at the back of the shoulder, as at the front.

FIG. 3.—Counter-irritation in abscess of the palm. Extent: hand and all forearm.

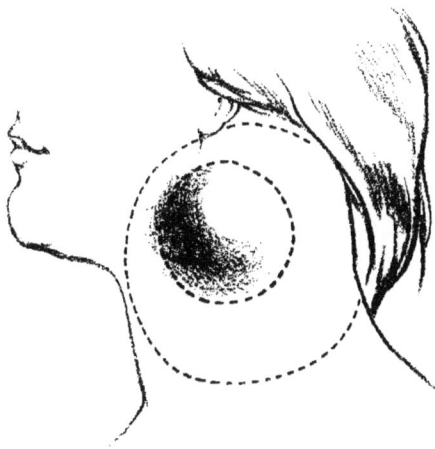

Fig 1

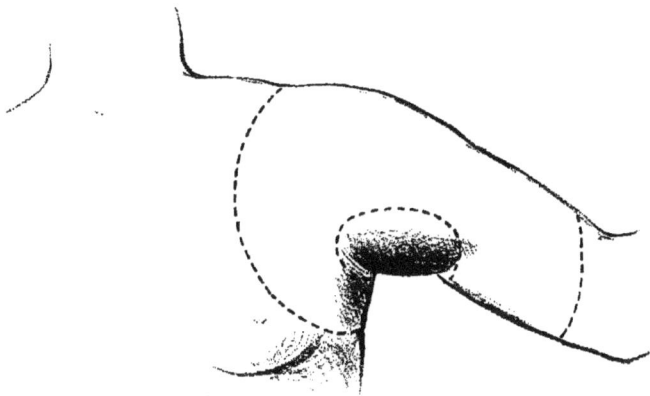

Fig 2

PLATE III.

Fig. 1.— Bubo or abscess in the groin. Counter-irritation may be carried over femoral artery. With mild counter-irritation the area should be larger, as in the next also.

Fig. 2.—Area of counter-irritation in gluteal abscess.

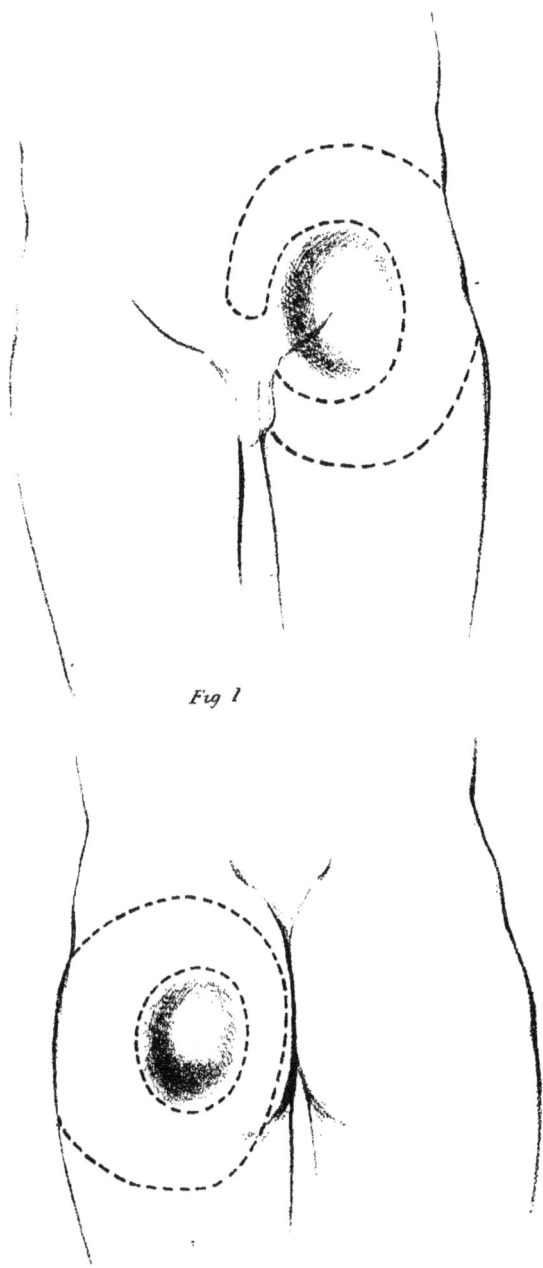

Fig 1

PLATE IV.

Fɪɢ. 1.—Area of counter-irritation in whitlow of thumb, namely, hand, fingers, and two-thirds of fore-arm, its whole circumference.

Fɪɢ. 2.—Counter-irritation, whole circumference of limb, in sloughing ulcer.

Fɪɢ. 3.—Counter-irritation in sloughing bubo (to be carried well over nates).

Fig 1

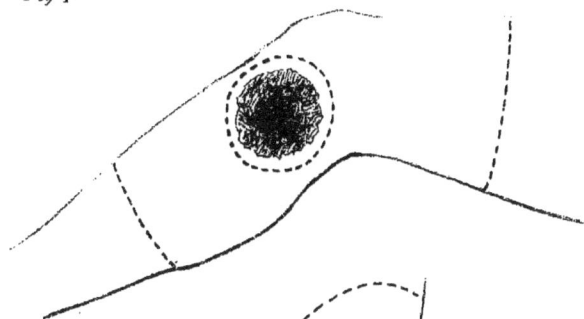

Fig 2

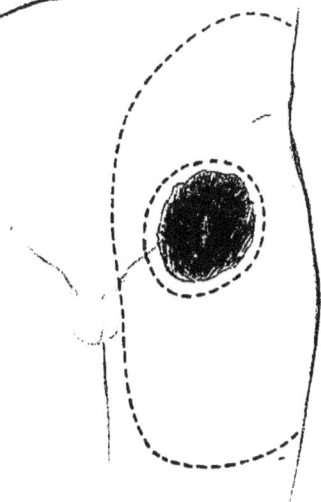

Fig 3

PLATE V.

Fig. 1.—Counter-irritation in carbuncle. A mild form should be more extensive, as in next fig. also.

Fig. 2.—Counter-irritation in a single boil, and in a cluster of boils.

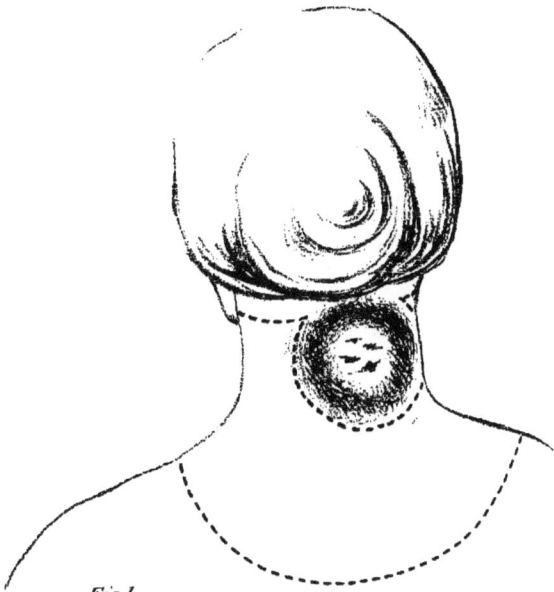

Fig 1

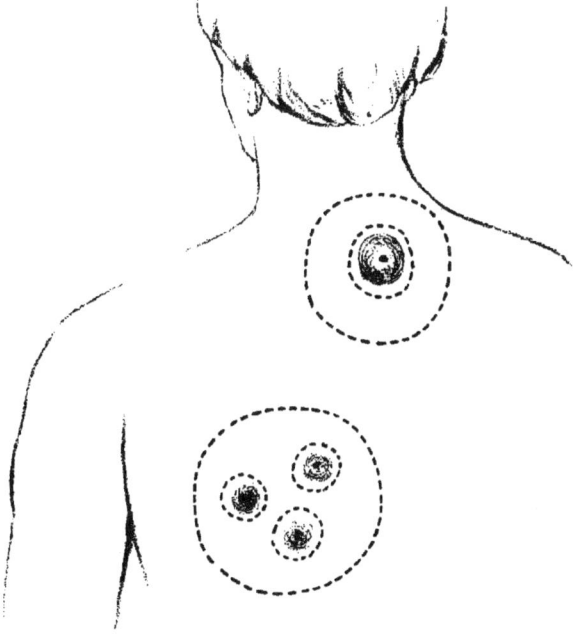

Fig 2

PLATE VI.

Fig. 1.—Counter-irritation in erysipelas and cellulitis, or diffused abscess. Acetum lyttæ may be used within the finely dotted lines.

Fig. 2.—Counter-irritation in erysipelas, cellulitis, or diffused abscess of leg. Iodine or arg. nit. ; acetum lyttæ in finely dotted lines.

Fig. 3.—Counter-irritation in chilblain.

Plate 6.

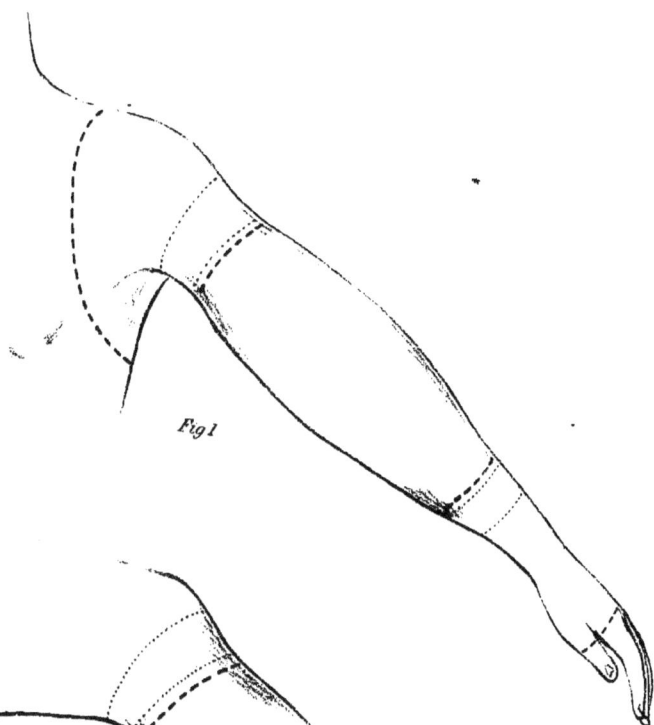

Fig 1

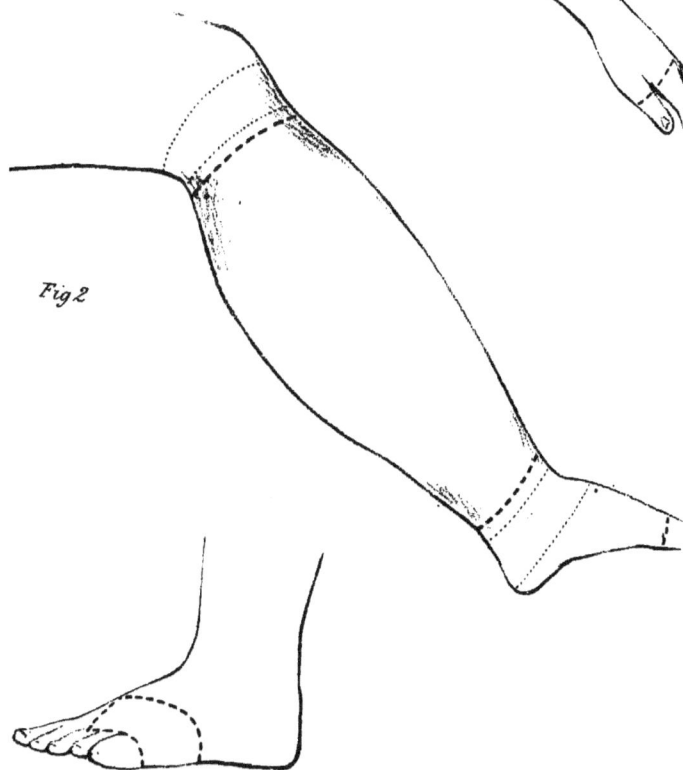

Fig 2

PLATE VII.

FIG. 1.—Counter-irritation in inflammatory diseases of shoulder, synovitis, ostitis, caries, &c. Immediately over the shoulder (within the finely dotted line), the application of (say) iodine should be less vigorous.

FIG. 2.—Area of counter-irritation in inflammations of elbow joint. In bone inflammations (the great majority), the counter-irritation should be more extensive and vigorous and protracted than in synovitis. This remark applies to all the joint diagrams.

FIG. 3.—Area of counter-irritation in wrist diseases. In advanced wrist disease see plate X, fig. 2.

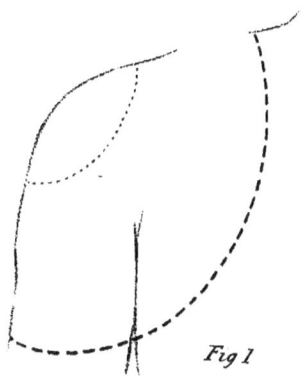

Fig 1

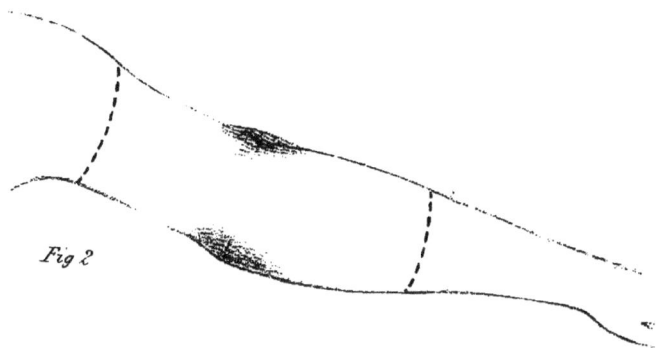

Fig 2

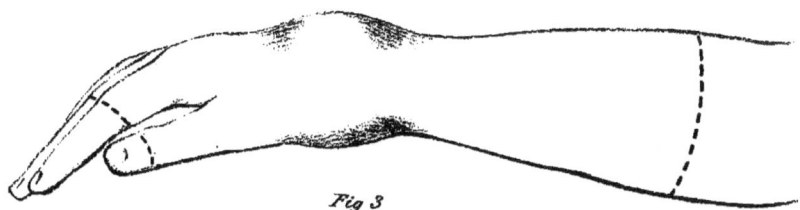

Fig 3

PLATE VIII.

Figs. 1 & 2.—Show extent of moderately vigorous (iodine pigment) counter-irritation in synovitis and early articular ostitis of knee and ankle.

Fig. 3.—Is intended to show use of actual cautery in severe articular ostitis of shoulder. One stripe is at the front, one at the back of the joint, and one midway between the two.

Plate 8

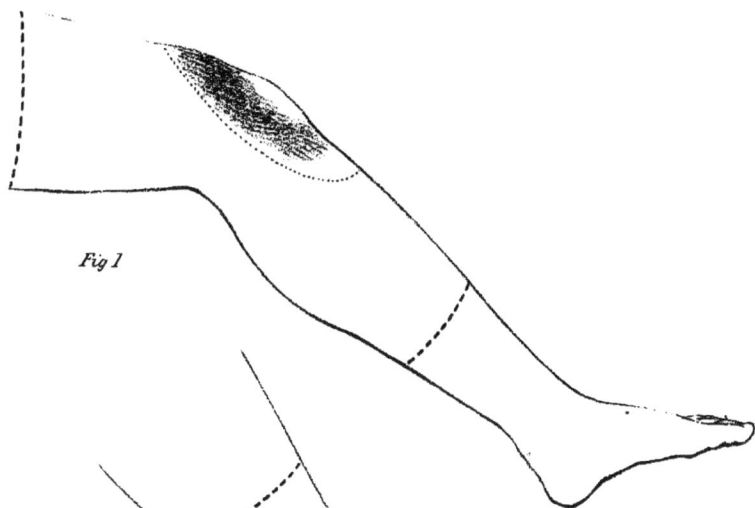

Fig 1

Fig 2

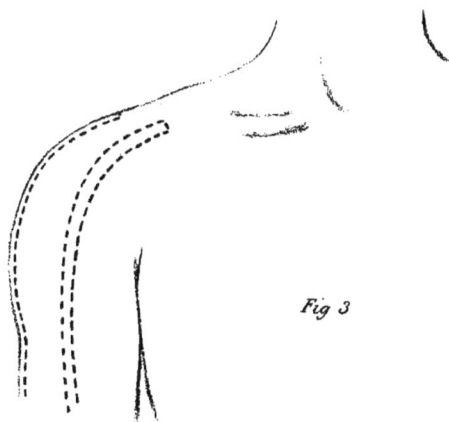

Fig 3

PLATE IX.

FIGS. 1, 2, 3, & 4.—Show mode of using actual cautery in advanced articular ostitis. In the hip a third stripe may be made over the trochanter. This treatment I have found strikingly successful in cases which have advanced to suppuration. I believe it entirely obviates the necessity for excision of joints, except (and for special reasons) the elbow, and it gives much better results. The actual cautery may be used with so much boldness (and with no drawbacks) that only a small residuum of cases should require amputation.

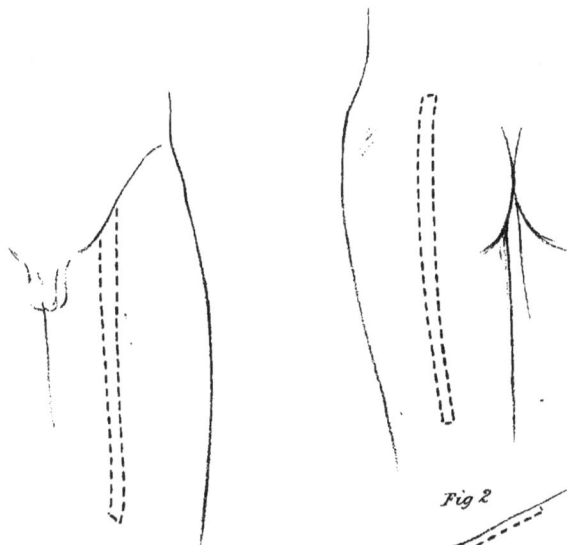

Fig 1

Fig 2

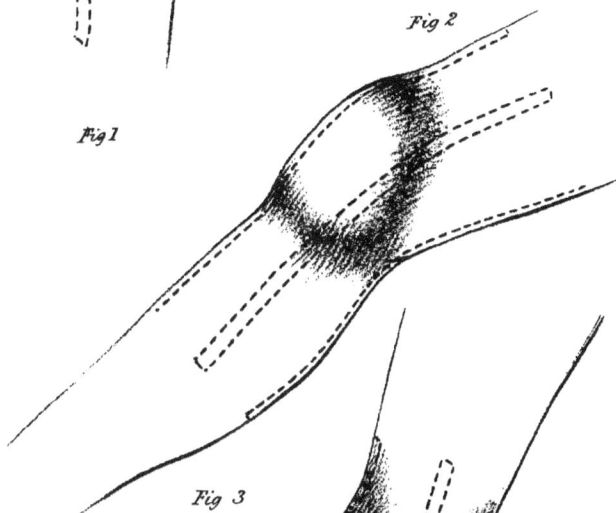

Fig 3

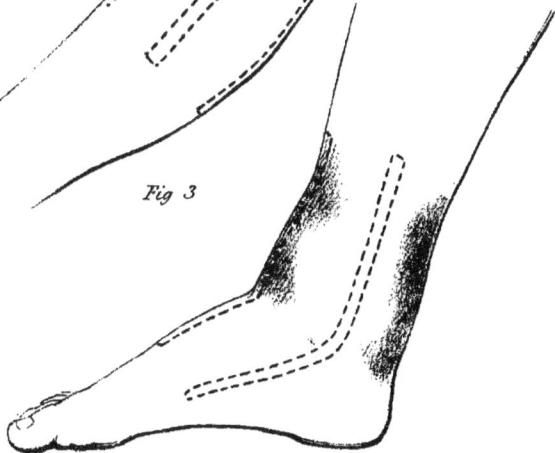

PLATE X.

FIGS. 1 & 2.—Show use of actual cautery in spinal caries and caries of wrist. See explanation of plate IX. I have seen spinal caries with abscess cured by this treatment. The commonly accepted doctrine is that in spinal abscess it is too late for the actual cautery, which means, that *in the severest inflammations we must not use our strongest remedies.* In wrist disease, however old, with or without sinuses, I have now for several years uniformly obtained a cure with the actual cautery, in broad long stripes, where necessary.

FIG. 3.—Shows extent of cantharides irritation in patellar bursitis.

FIG. 4.—Shows localities and extent of iodine irritation in inflammation of sheath of tibialis posticus.

Plate 10.

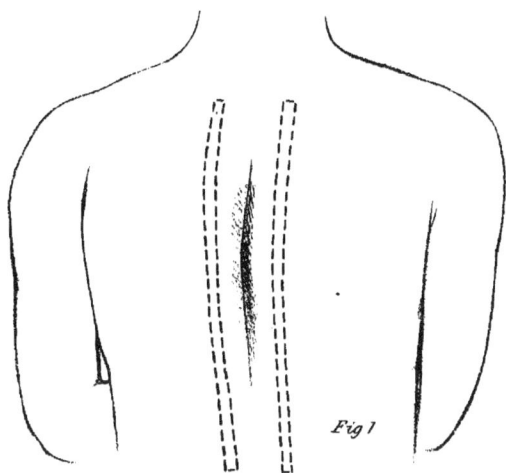

Fig 1

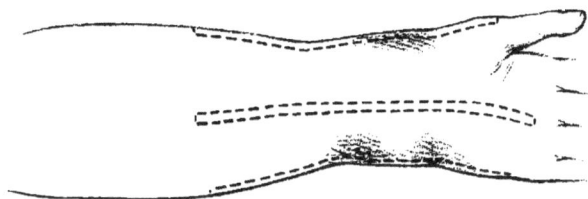

Fig 2

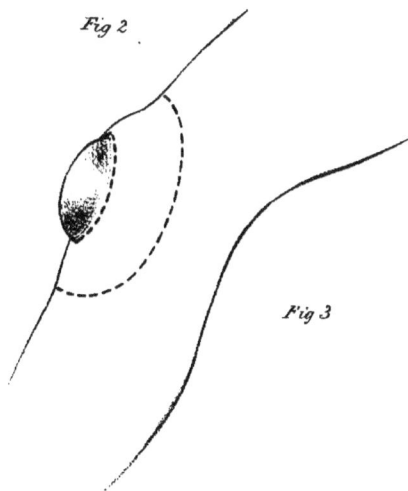

Fig 3

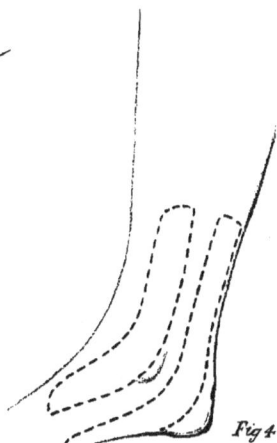

Fig 4

PLATE XI.

FIGS. 1, 2, & 3.—Show area and mode of applying counter-irritation (iodine) in nodes of tibia, ulna, and clavicle. Not only is the subsidence of the swelling extremely rapid as a rule, but, and this is an invaluable result of counter-irritation, the relief of pain is instantaneous.

Plate II.

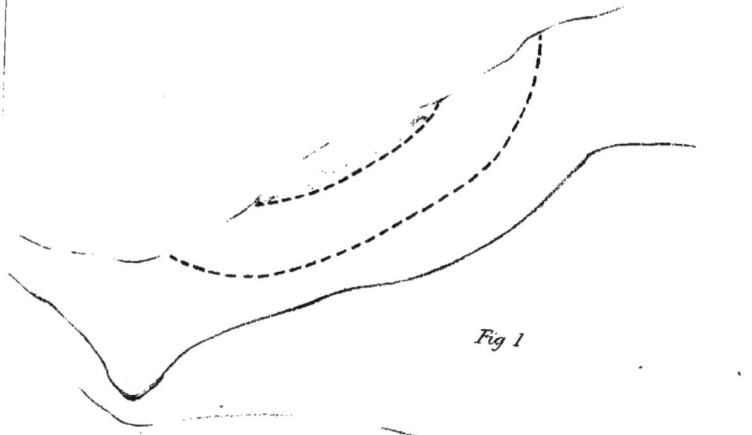

Fig 1

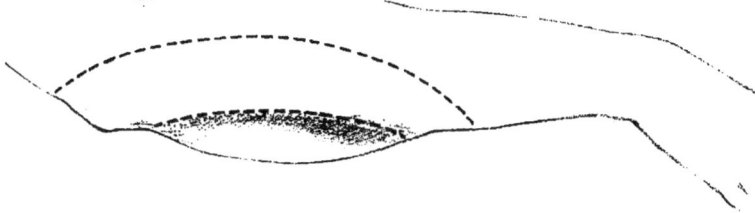

Fig 2

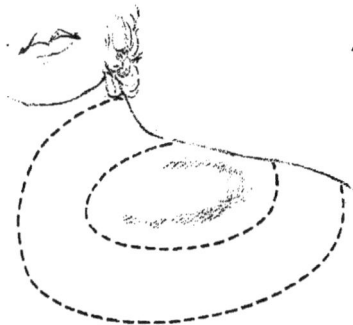

Fig 3

PLATE XII.

FIG. 1.—Shows mode of using shot-pressure in enlarged glands at intervals during the day, according to leisure and patience of patient.

FIGS. 2 & 3.—Show counter-irritation (iodine) in enlarged glands. See Mr. Hickinbotham's case, where, in a young man, after a trial of every known treatment by numerous surgeons during a period of three years, the treatment by adjacent counter-irritation completely succeeded in three weeks.

Plate 12

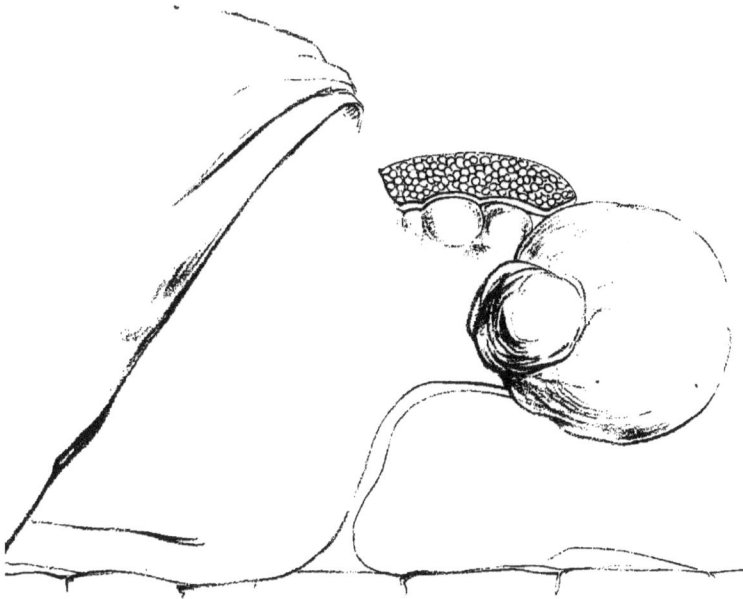

Fig 1

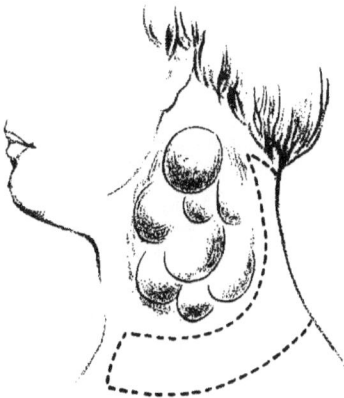

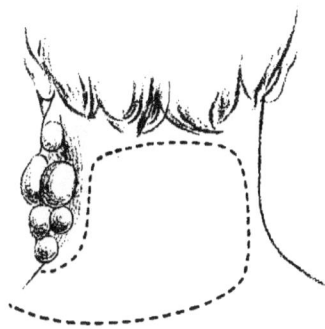

Fig 2.   *Fig 3*

PLATE XIII.

Fɪɢ. 1.—Suggests modes of using counter-irritation in inflammation of ear, eye, nose, mouth, tongue, tonsils, pharynx, and larynx. In painful abscess of tympanal cavity or meatus of ear, the counter-irritation may, with great benefit, be carried below and in front of the ear.

Fɪɢ. 2.—Shows extent and locality of counter-irritation in bronchocele. Elastic pressure should be combined with this.

Fɪɢ. 3.—Counter-irritation in inflammations of the tongue. I have known this very satisfactory, even in troublesome specific ulcers.

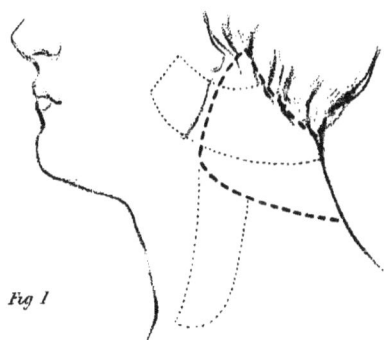

Fig 1

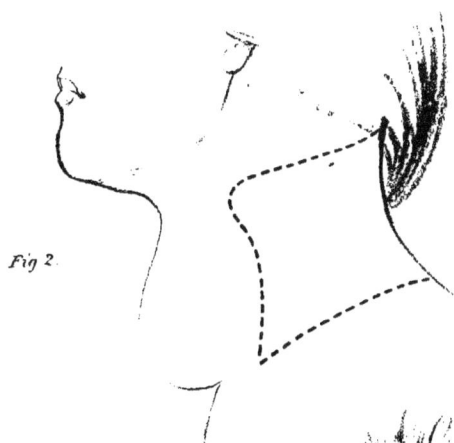

Fig 2

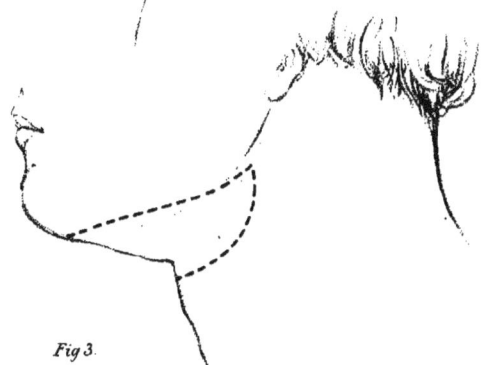

Fig 3

PLATE XIV.

FIG. 1.—Shows mode of using counter-irritation (acetum lyttæ in severest, iodine liniment in less severe cases) in purulent inflammations, especially gonorrhœal, of conjunctiva. Where counter-irritation must be repeated or kept up, two horseshoe shapes, one within the other, are useful.

FIG. 2.—Shows sharp counter-irritation in the more acute laryngeal inflammations.

FIG. 3.—A milder counter-irritation in the less acute inflammations.

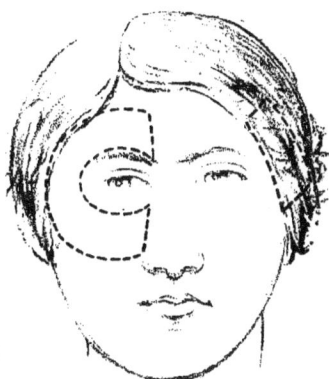

Fig 1

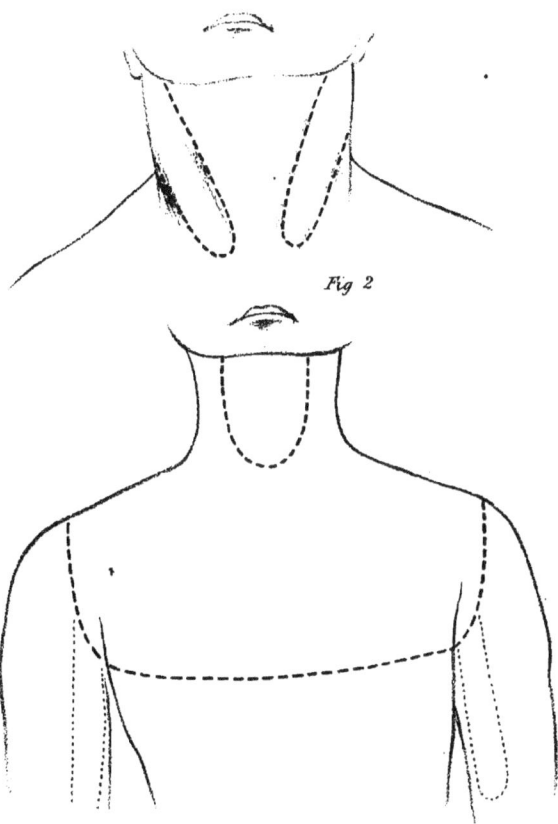

Fig 2

PLATE XV.

FIG. 1.—Shows sharp counter-irritation (acetum lyttæ, or repeated iodine liniment) in inflammations of a hernial sac. In the case from which the diagram was taken, excessive local inflammation, and general symptoms of intestinal obstruction, were removed in twenty-four hours by an efficient circle of acetum lyttæ.

FIG. 2.—Shows counter-irritation (acetum lyttæ, or iodine liniment freely) in the acuter pelvic inflammations, especially gonorrhœa, where there is no suspicion of stricture, as in first attacks. In acute orchitis, the scrotum should also be painted with arg. nit. The worst cases may be cured in twenty-four hours.

FIG. 3.—Shows a disc of milder counter-irritation for gonorrhœa, gleet, and less acute pelvic inflammations, inflammations of urethra, prostate, bladder, vagina, uterus, pelvic cellular tissue.

FIG. 4.—Counter-irritation in inflammation and abscess of female breast. With a mild iodine paint the area should be larger.

FIG. 5.—A disc of counter-irritation in uterine inflammations.

Plate 15

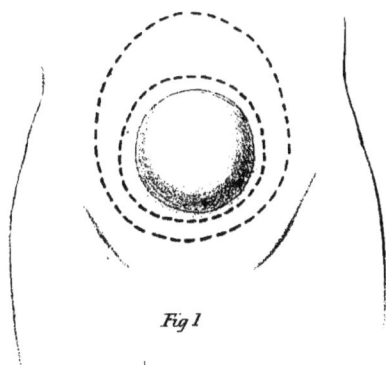

Fig 1

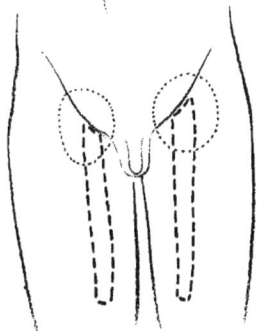

Fig 2

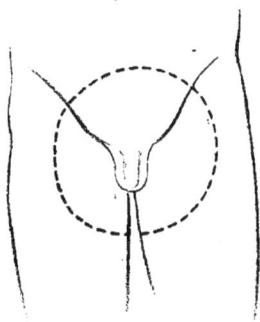

Fig 3

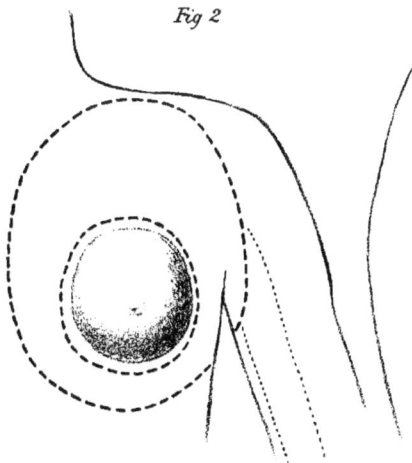

Fig 4

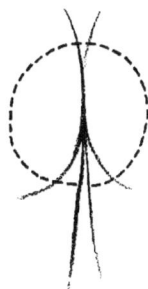

Fig 5.

THE TREATMENT OF

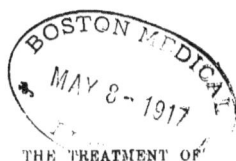

SURGICAL INFLAMMATIONS.

I.

FREQUENCY OF SURGICAL INFLAMMATIONS—VARIETIES OF TREATMENT.

THE great majority of diseases which the surgeon treats are of an inflammatory character. They form a large proportion of any surgical nomenclature, they form a still larger proportion of the cases in actual practice. If we look at the diseases of the cutaneous structures, we shall find that the frequency of cancer, nævus, and warts, is slight, contrasted with the frequency of abscess, boil, carbuncle, onychia, paronychia, erysipelas, and the ulcers. In the bones, again, periostitis, acute and chronic, ostitis, necrosis, caries, abscess, sclerosis, preponderate numerically over rickets, mollities, tumours, and cancer. It is so, more or less, with all the organs, the diseases of which fall to the lot of the surgeon to treat.

I have taken some pains to ascertain the percentage of inflammatory and non-inflammatory diseases in

B

the out and in-patients of the Queen's Hospital, between two fixed dates. In round numbers, of the out-patients five-sixths of the diseases were inflammatory, of the in-patients two-thirds were inflammatory. The disparity is probably due to the entrance into hospital of cases specially for operations, as cancers, tumours, herniæ, vesical calculi, and others. These conclusions are drawn from 3,913 out-patients and 1,179 in-patients.

Excluding for the present specific inflammations, as the syphilitic and rheumatic, no one contends that the inflammatory action is a different process in the different localities, organs, and tissues. How comes it, then, that there are numberless varieties of treatment for inflammatory diseases ? How comes it that there is one list of treatments (if I may say so for brevity) for an inflamed urethra, another list of treatments for inflamed prostate, another for inflamed bladder, another for inflamed testis, one list for inflamed tympanal membrane, another for inflamed tympanal cavity, one for inflamed tongue, another for inflamed tonsils, another for inflamed larynx, another for inflamed bronchus ? How is it that there is one treatment for the beginning of inflammation, another for the middle, another for the end ? In systems of surgery the chapter on the treatment of inflammation gives a long list of remedies. In the subsequent

chapters these are forgotten and endless new ones
introduced.

Multitudes of remedies probably date from times
antecedent to any definite pathological knowledge, and
men have not the industry to inquire into their value,
or the courage to disavow them.   In these pages,
while I freely recognize specific and modifying influences in certain inflammations, I shall assume that
the inflammatory process is one, in whatever part and
under whatever circumstances it may appear.   Its
causes, modes of progress, degrees of activity, and
results are many and various, but the process is the
same; the phenomena in the main are also the same,
and the same also are the changes in the tissue
elements as discovered by the microscope.

Inflammation, then, being *one* disease, why should
we not have *one* treatment.   Having discovered the
best remedies in some inflammations, why not use
them in all?   If a mustard plaster will cure a slight
bronchitis more quickly than any drug or any other
remedy, will not a stronger counter-irritant be the
best remedy for a severer bronchitis?   More than
this, if one inflammation is more quickly than by any
other means subdued by counter-irritation, why not
try the same remedy in other inflammations, why not
try whether a mustard plaster, or other counter-
irritant, around a carbuncle or an abscess or a phage-

dæna will not, with equal superiority, control their progress? If pressure will remove a chronic synovitis more quickly than any drug, or any other remedy, except counter-irritation, why not employ pressure, gently in acute, firmly in chronic cases, in every accessible inflammation?

II.

THE OBJECTS AND DEPARTMENTS OF SURGICAL TREATMENT.

THE treatment of every surgical disease may be classed under one, or more, of four heads. It is either, 1, of the nature of an operation, as amputation and many others; or it is, 2, mechanical, as in spinal curve, stricture, most injuries, and many diseases; or, 3, it is directed to the removal of inflammation, as is the case in the majority of diseases; or, 4, it is directed to the cure of conditions of obscure pathology, such as tetanus, hysteria, pyæmia, hydrophobia, and a few others. Here the treatment is chiefly by drugs, but no one contends that any considerable potency can be claimed for any particular drug or drugs in the treatment of these diseases.

III.

THE PRINCIPLES WHICH SHOULD GOVERN THE TREATMENT OF INFLAMMATION.

THE essence or the nature of the inflammatory process is unknown, and hence it is fruitless to attempt to extinguish it by any single drug or internal remedy. But the inflammatory process is attended by certain phenomena, which are so regular that they may be said to be conditions which are essential to its existence. If there be one condition which is really essential, its removal would involve the removal of the inflammation itself. There are, probably, a few conditions which can be more or less removed. The difficulty of removing them completely is the practical difficulty in the treatment of inflammation. An illustration of the principle here put forth may be obtained from life. The essence or nature of vitality is unknown, but we know the conditions under which life exists, and if we desire to put an end to life we have simply to remove an essential condition—such, for instance, as the supply of food.

What are the principal conditions which are needed for the prosperous beginning and progress of an inflammation? An inflamed part must have increased space —more room for swelling, it must have more blood, it

is all the better for perpetual unrest.   There must be
no other inflammation near to interfere with it—two
inflammations do not flourish together.   It must have
a cause.

An apparently essential condition to the existence
of inflammation is *increased space*—pathological action,
multiplication of tissue germs (or exudation), more
blood, all require more space.   If it were possible to
keep any part of the body within its physiological
precincts it could not inflame.   If by *pressure* we
could in some degree restore an inflamed part to the
area of health, the inflammation would be removed to
a proportionate degree.

Another essential condition of inflammation is increased quantity of blood or "ministering" fluid.   No
part of the body can be inflamed if the "health
quantity" only of blood be present.   Local diminution
(*general* loss of blood means loss of repairing power)
can be effected by pressure and elevation where practicable, and especially by exciting a *second* inflammation over another, say the next vascular territory.
Small local depletions and pressure upon or occlusion
of the feeding artery act on the same principle.

Every inflammation must have a cause, and this
suggests an obvious principle of treatment never to be
neglected.   Where there is a mechanical cause for
inflammation, as when a stone in the bladder causes

cystitis, or a stricture maintains a gleet, or a small meatus or a phymosis produces a cystitis in a child, or a ragged nail produces granulations, or a shrivelled nail keeps up an onychia, or a foreign body keeps up suppuration, treatment directed to the inflammation only is of course altogether futile. Unfortunately, in the greater number of inflammations the cause is unknown, or so little known, or so irremovable, that the treatment must be based on other grounds. In many the cause is transitory, and the result only needs consideration.

A condition which keeps up and aggravates inflammatory action is *unrest*. Movement is as mischievous to an inflamed microscopic tissue element as it is to an inflamed joint or eye. Rest is a remedy so necessary as a foundation to all other treatment, that it can only be regarded as an evidence of the slow progress of therapeutics that it should have been left to a surgeon of the nineteenth century (Mr. Hilton) to enforce its universal need.

Another condition which appears to me very suggestive as regards treatment in this: where inflammation prospers there is no other inflammation present. If a second inflammation, say an abscess, arises, the first shrinks or disappears. There is no inflammation in which this condition may not be removed by establishing another and counter-inflammation. But the counter-inflammation must be strictly a second, and never an aggravation of the first inflammation—as counter-irritation over the thin parietes of the cranium, or thorax, or abdomen, or knee joint, or subcutaneous abscess, may readily be, if deeply acting irritants are used.

IV.

THE BEST REMEDIES FOR INFLAMMATION—THEIR SELECTION.

IT is scarcely necessary to say that the grand basis of the selection of remedies must be that of success— success which must be judged by the results of experience. How to select these: It has for a long time appeared to me that if in any inflammation there are any remedies which are quicker in action than others, *such* remedies should be tried in all inflammations. It seems so natural to suppose that the remedies which are the best in some inflammations, should be the best in all, that they should at least be proved to be inferior before others are tried. I have found an uniform plan of treatment the most successful.

To logical minds no remedy is the worse if it is capable of intelligent explanation. To such minds, if any select class of remedies are not only in practice the best, but are also the most consistent with physiological and pathological laws, their value is not diminished. If two minds, or one impartial mind at two different periods, started on two different routes in search of the best remedies, and both hit on identically the same remedies, the grounds for using such remedies would not be weakened. I have attempted to do this. I sought with a single desire for truth (and the usefulness of truth) to discover the best remedies for inflammation on theoretical grounds. In doing this I necessarily passed in review the pathology of inflammation, and as necessarily came to the conclusion that in all disease, in life, in death, in science, we only know phenomena. So in inflammation. We know nothing except its phenomena—phenomena pathological and clinical. If these phenomena were constant they were essential; if essential and capable of removal, their removal would remove the process of which they were the evidence. Such phenomena appeared to me to be the assumption of increased space, the presence of a larger quantity of blood, the absence of a second inflammation, the absence of rest, often the presence of a mechanical cause. How, then, could I prevent or diminish the assumption of space? I knew of no means except pressure, not "pressure with a will," as a hasty reviewer has said, but pressure carefully adapted to the acuteness, or chronicity,

c

or sensitiveness, of the part. [I would say here, parenthetically, that I shall be thankful to have it shewn that the principles here enunciated are erroneous; that I have nothing to do with the conditions or the phenomena, and ought not to try to remove them; or if the principles be admitted, I shall be still more thankful to be shewn (as is very possible) that the conditions or phenomena of inflammation can be removed by better methods than are here proposed.]

Counter-irritation, by calling the " ministering" fluid to another inflammation, might assist, as also might elevation or local venæsection. How could I diminish the quantity of blood? I knew no better means than counter-irritation, and pressure, and elevation where practicable. How could I supply a second inflammation that should tend to subdue the first? Readily by counter-irritation, and in definite degrees. How I could remove unrest or causal influences, were, of course, questions of detail in given cases. I then, and entirely without bias I believe, endeavoured to discover the most rapid cures of inflammation, and under what circumstances, and by what means. I had repeatedly known a smart catarrhal bronchitis effectually relieved in sixty minutes by the application of a large mustard plaster for fifteen minutes. I had several times seen an acute synovitis of the knee completely disappear in twenty-four hours after the application of a strong solution of nitrate of silver. These were the most rapid instances of cure of inflammatory disease that I knew.   I knew no other local or internal remedy that was so rapid.   Cold, locally, is in most cases very painful, and I have never seen any rapid result from its use.   It would seem that directly applied to an inflamed part it ought to diminish vascular action, because it is the one agent which depresses all action of whatever kind.   The probable explanation of its comparatively practical inutility is that its action is only superficial.   To cool or freeze the skin merely over an inflammation, even where the skin is involved, tends to aggravate the inflammation.   Emetics and purgatives often do good, and unquestionably act as counter-irritants, one to the stomach and the other to the intestinal canal; but in most surgical inflammations at least there are localities in which counter-irritants are more rapid and efficacious.   I knew no drug that specially acted on a local inflammation.   In specific inflammations there are certainly drugs of great utility, although of no marked rapidity of action.

Next to counter-irritation, the most striking cases of rapid removal of inflammation which I had seen were those in which pressure had been used.   I had seen an inflammatorily thickened synovial membrane quickly thinned by pressure.   I had seen a shot-mattress over

a bubo, or an abscess, or a carbuncle, invariably
accelerate the disappearance of those diseases so
strikingly, that the only remedy I could compare to it
was counter-irritation.   I had so constantly seen
inflammation get well, when rested, which would not
get well before, that I felt no doubt of the universal
necessity of this remedy.

I resolved, therefore, having taken every care to
remove the cause, where it could be removed, and
having also secured rest by appropriate means, to try
counter-irritation in every inflammation, combined
with pressure in accessible inflammations, and elevation where practicable.   The results were, in many
cases, marvellous, often extensive inflammations were
removed in a few days, and, in nearly all cases, in a
much shorter period than is usual with other treatment.
In a very few cases little or no benefit followed ; these
were either examples of strong dyscrasiæ, as syphilis,
or struma—the dyscrasic influence indeed predomi-
nating over the inflammatory, or some undiscovered
mechanical cause existed, as in the case of a gonorrhœa
with a previous stricture.

The remedies, then, which appeared most reliable in
inflammatory diseases were counter-irritation, pressure, rest, elevation, and removal of cause.   My rule
of treatment was this in every inflammation (discard-
ing, for the time, authorities, books, teachers, and

friends), how and where can I best employ counter-irritation? How best, and to what degree, can I use pressure? How can I best secure rest? Can I elevate the part? Have I taken every care to remove the cause?

Internal treatment did not appear unimportant; on the contrary, in specific inflammation I firmly believed as the result of daily observation, in the great utility of mercury and iodide of potassium. In all inflammations I gave iron. Often, as the great "rester," I found opium invaluable.

V.

COUNTER-IRRITATION.

I have already referred to the rapid effects of counter-irritation in acute inflammations, as orchitis, synovitis, and others, although its mode of action has been represented as incapable of explanation (and indeed as a consequence that it has no action), yet I venture to say that we possess no remedy whatever whose action is as well understood. Purgatives are allowed to be of use in certain cerebral diseases, but who will doubt that they act as irritants to the intestinal canal, as counter-irritants to the brain. I have observed in myself more than once the singular effects of what I

may call nature's counter-irritation. I have gone into
the lecture theatre with a most severe cold, the ocular
and nasal membranes secreting most copiously and
constantly. With the cerebral excitement of an extem-
pore lecture every symptom of cold has disappeared,
the eyes and nose remaining in perfect comfort until
a longer or shorter period after lecture.. Was not the
increased vascularity and excitement of the brain an
undoubted counter-irritation to the nasal and ocular
mucous membranes ? By what action is it, if not by
counter-irritation, that a smart gonorrhœal discharge
completely disappears when acute orchitis sets in ?
On this subject I might bring more facts and illustrations. They are unnecessary.

Neither am I at all anxious to give a rational explanation of counter-irritation. I cannot explain
chloroform, but I would not have my child's finger
nail taken off without it. But, as I have already said,
we can understand this better than any other remedy.
It certainly takes away blood from an inflamed part.
This may be watched sometimes. In an erysipelas, I,
and half a dozen with me, had seen the colour distinctly fade, and the swelling distinctly subside in
twenty minutes after the establishment of a broad
ring of counter-irritation with iodine liniment. I and
others have heard a loudly noisy respiration in œdema
of the vocal chords become quite calm in fifteen

minutes after smart counter-irritation over the carotid and brachial arteries. There is probably also some reflex influence of the spinal cord. We know at any rate that there is an intimate inter-communication between the cerebro-spinal and vaso-motor systems. Mr. Simon makes a suggestion touching the action of counter-irritation, which I think has much force : a certain amount of textural force being put into operation in a given locality, some—much or little, may on the principle of the correlation of forces be transferred to another and a safer locality.

Seeing that counter-irritation in some inflammations is so rapidly beneficial, I have often asked myself why it is not more frequently used. I can only find one answer —there are no clear or acknowledged principles by which to determine its localities. Counter-irritation as often used is simply direct irritation. It is possibly so in pleurisy, especially in the early stage. The same may be said of the abdomen, the cranium, of thinly covered bubo or joint. But counter-irritation over another, more or less independent vascular territory, tends to arrest the original inflammation, wherever it may be. I have little doubt that in the earlier stages of pleurisy a blister covering the inner side of the arm would rapidly subdue inflammatory action. I do not refer, of course, to pleurisy the result of tubercular or foreign-body irritation.

It is not contended that the localities for counter-irritation suggested here are the best. Future experience will certainly bring improvement. The principle laid down, however, is clear and definite, and in practice has been shown to give results of an eminently striking and satisfactory character. *Counter-irritation should be established over the next, or another, or an independent vascular trunk or territory.* In intra-cranial inflammation, counter-irritation should be excited over the branches of the external carotid—over the neck generally, if a superficial irritant be used, over the mastoid process or back of the neck, or both, if a deeply acting irritant be used. In inflammations of the wall of the thorax, as pleurisy internally, and abscess, or carbuncle, or erysipelas, or mammary inflammations externally, the counter-irritation should be over the brachial artery if deeply acting irritants are used, or over the thorax itself if superficial agents are used. In abdominal and pelvic inflammations, deep counter-irritation should be excited over the femoral arteries. In all these instances the inflammation and the counter-irritation may be reversed; in inflammations of the upper extremity, or the thigh, irritation may be excited over the thorax or abdomen respectively. Although only a detail, I believe there is some advantage in putting stripes of counter-irritation over the arteries. The

arteries are in sheltered positions, larger branches, of
course, lie nearer the trunk, and as the larger nerve
trunks lie mostly with the arteries, the nerve influence,
if any, will not be lessened by the locality and configuration referred to.

It will be seen already that counter-irritants may be
roughly classed under two heads—the deep and the
superficial. In the agents used for artificially producing
inflammation there is room for much improvement,
especially is this so as regards certainty, definiteness of
degree, and painlessness of action.   I have chiefly
used iodine, nitrate of silver, and cantharides.   Iodine
is very convenient because it can be used in so many
degrees of strength from the tincture to the strongly
acting liniment, intervening degrees of strength being
obtained by mixing in different degrees the weak
tincture and the strong liniment.   The mixture may
be conveniently designated a " pigment."   Strong
solutions of nitrate of silver are also excellent irritants ;
they are apparently a little more painful than iodine.
The acetum lyttæ (made with glacial acid) is an
excellent application for less extensive surfaces.   It is
painful, but only for a very short time.   Where
counter-irritation requires to be maintained for a length
of time and over a large surface, an iodine pigment
is perhaps the most convenient agent.   If the artificial
inflammation of nitrate of silver or cantharides

D

requires to be prolonged, it may be by new patches, or stripes, or circles, or the use of cantharides ointment after cantharides. Near the urinary organs arg. nit. is preferable to cantharides. In children counter-irritation should be excited with care, especially in the very young. Slighter agents over a larger surface are better than the deeper irritants. No important irritation should be excited in children without the kindly aid of chloroform, which with them is so safe and pleasant. In addition to the agents referred to, mustard, turpentine, hot water, and many others may be occasionally used.

A very simple and successful mode of effecting counter-irritation, especially where superficial irritants are used, is to adopt the form of circles, or zones, or horse-shoes, or crescents. Where a deep irritant is used the circle must be narrower. Or a stripe of deep irritation may be combined with the zone of superficial irritation; for instance, in abscess of the mamma, a zone of iodine pigment, permitting local pressure, may be combined with a stripe of acetum lyttæ over the brachial artery.

What is the use of counter-irritation in specific inflammation, such as syphilis? Its utility is great, but only as subsidiary to specific internal remedies, such as mercury. In a destructive syphilitic inflammation a sharp and limited second inflammation is probably best effected by means of the glacial acetum lyttæ. In rheumatism and struma destructive (sloughing) inflammations are not often met with, and there is more leisure for the selection and use of remedies. In strumous diseases of the bones, and glands, and skin, a counter-inflammation is the most rapid (and that may be slow) treatment, but it should be associated with other local measures and careful general treatment.

The great advantages of counter-irritation, independently of its efficiency as a remedy, are its applicability to every inflammation, and its capability of being estimated in extent and intensity.

A most striking feature of counter-irritation is its immediate and certain relief of pain.

VI.

PRESSURE.

I HAVE already remarked that if any given part of the body, including each individual tissue element, could be restrained within its physiological area it could not inflame   Of this fact nature herself furnishes a striking example. In so-called acute orchitis the inflammation is really in the epididymis. The inflammation travels from the urethra to the epididymis by an unbroken surface, and by the same surface might naturally be expected to enter the testis. That it should suddenly break off here, and expend itself in the adjacent connective tissue, can only be explained on the ground that there is no room within the tunica albuginea for sudden acute inflammatory action. In larger circumscribed cavities, such as the skull, inflammatory swelling is effected by the compression of the surrounding parts. In the eyeball, too, where acute inflammation occurs, though rarely, the varities and number of structures permit of the easy alteration of the space occupied by one tissue in relation to the rest.

It has been objected that the origin and progress of inflammation should be difficult in bones because of the natural pressure of the tissue elements. The probability is that inflammation, especially the acute, is very difficult to excite in bones. It has been frequently remarked that in the young, in whom bone inflammations mostly occur, and nearly all acute inflammations, that the physiological changes in bone are so rapid and so various that it takes little to pass over the line which separates physiological from pathological action. Notwithstanding this indisputable fact, we find that acute inflammation of bone is extremely rare compared with inflammation of the the soft tissues. If we were to take the first hundred men we met, probably not one would have escaped a boil, or carbuncle, or abscess, or whitlow, and possibly not one would have had an osteal inflammation.

Pressure probably acts by diminishing the assumption of space, and by diminishing the quantity of blood in a part. It tends also to secure some degree of rest to the inflamed part.

Like every other important remedy it requires great care in its use. It should not give rise to pain. It should be gentle in acute and early stages of inflammation, it should be especially gentle where destruction of tissue is threatened, it should be firmer in chronic and the later stages of inflammation. Neither pressure, nor counter-irritation is here advanced as a nostrum that any one may use under any circumstances. On the contrary, they require judgment and experience as to the method, and the extent, and agencies which shall be adopted.

There is probably not a single accessible acute inflammatory disease in which some degree of pressure may not be used. The *pressure* and the *heat* (counter-irritation) of a poultice are not only bearable but " comforting " in the acutest and tenderest inflammations. Or is there any virtue in linseed meal, independently of its being a medium for pressure or heat?

Shot-mattresses, from the lightest to the heaviest, strapping, easy or firm, bandages, slacker or tighter,

are the simplest means, by which pressure may be
effected. A shot-mattress over, or a bandage around, a
poultice materially augments its benefit, even in acute
inflammations. Chronic inflammatory products, say
in the breast, testis, cutaneous structures, knee-joint,
&c., &c., are more rapidly and effectively dispersed
by carefully adjusted pressure than by any other
treatment. I believe the future will bring improved
methods of exercising and controlling pressure.

VII.

REST, ELEVATION, REMOVAL OF CAUSE, GENERAL TREAT-
MENT, DRUGS, COMBINATION OF REMEDIES, RESULTS.

As I come now to consider those remedies, to the
utility of which all readers will agree, I may be very
brief—brief for this reason only, and by no means on
account of their unimportance. There is perhaps no
inflammation, unless it be chronic rheumatic arthritis
(which Dr. Todd said was not an inflammation, and
which certainly is not an ordinary inflammation), in
which too great pains can be taken to secure rest.

Elevation of inflamed parts is more frequently
acknowledged to be useful than it is adopted. In the
limbs it is often necessary simply to relieve pain, for
the tension of an inflamed foot, or leg, or hand, or

fore-arm, is greatly diminished if the veins are emptied by elevation, at the same time that a barrier to local circulation is removed. In elevating the limbs care should be taken, where there is no good reason to the contrary, that the joints are somewhat flexed, as they always are when in repose.

There is one kind of elevation to which I shall draw attention here, because of its great utility, in my opinion at least. It is that of elevation of the pelvis in inflammatory diseases of the pelvic organs, by placing a flat pillow under the nates. It is especially useful in hip disease, because, besides assisting the local circulation, it tends to overcome, by the weight of the limb itself, the tendency to flexion which is often difficult to counteract, and which, if it be allowed to remain, causes a little limp in an otherwise cured case.

As regards the removal of the cause of inflammation it is impossible to exaggerate its importance, and if the treatment advocated here, or any treatment diverted attention from this cardinal principle it would be an evil and not a benefit.

What is the province of drugs in simple or non-specific inflammations? I believe that many drugs that are used are of no benefit whatever. We should certainly see that the different organs are acting healthily, as the stomach, intestinal canal, liver, kidneys, and skin, but we should secure this rather by habits and diet than by drugs—occasionally drugs are of undoubted use.   On physiological and pathological grounds, still more because in actual cases I have found most benefit from it, I give iron in all or nearly all inflammatory diseases.   I also give it with mercury in primary and secondary syphilis, and with iodide of potassium in tertiary syphilis.   I give it also in struma and rheumatism. I prefer usually the ammonio-citrate, with a little alkali, well diluted, and before food.

The general hygienic and dietetic treatment will vary with the severity of the inflammation.   A milk diet I regard as of the first importance.   Animal food should be added where it can, and in such quantity as can be digested.   In the great majority of surgical inflammations stimulants are unnecessary, but where there is much shock, or exhaustion, or profuse discharge, they should be given, and given often with a prodigal hand.

All moderately severe inflammations require rest in bed.   Sometimes, even in chronic and slight, or at least not invaliding, inflammations, rest in bed will secure recovery when other means fail.   Ventilation is certainly important, but it is less (and I purposely speak strongly on this point) important than warmth. Warmth of the whole body is to me a corollary to counter-irritation.   To chill the whole body and put a

blister on one part of it, is a practical contradiction.   I
believe the present mania for open windows in all
weathers is answerable for many a death.

The different remedies which are here believed to be
best, where it is practicable, should be used together.
In many inflammations, as prostatitis, cystitis, urethritis, &c., pressure of course cannot be used.   Where
they can be used in combination they may require to be
used in very unequal proportion; in an acute inflammation, an abscess, or an erysipelas, counter-irritation
should be the principal treatment in the earlier stages
and pressure in the later.

A very convenient and successful combination is
that of pressure to the inflammation, and counter-
irritation over the next artery.   In inflammations of
the mammary gland, or the testis, the inflamed organ
may be compressed, and the brachial or the femoral
" lines " be irritated.

It requires that all these remedies should be used
with thoroughness and care.   It is easy to carry out
the treatment in a perfunctory manner, or adopt it
in some case where there is some mechanical or marked
specific cause for inflammation, and express disappoint-
ment at the result.

Touching the results of the treatment which I now
advance, after an experience of three or four years, I
point to the cases and to the careful trials and observa-

E

tions of others. The cases are made as brief as possible, as they are simply intended to faithfully illustrate the nature of the disease and the duration of the treatment. Nearly all the cases were hospital cases; they were seen by many watchers, and are entered on the hospital registers.

I contend that, judging from results solely, the treatment is much more successful than any other. It secures recovery often in a mere fraction of the time required by ordinary treatment. Where the treatment is long, it is, nevertheless, the shortest and best. Where it fails, as in this world every kind of treatment must, it has at least given a better chance (experience justifies me in saying this) than any other treatment could have given. It cannot undo structural changes, but it can often arrest these, or reduce them to their narrowest limits. If sloughing has commenced it will, by rapidly removing the inflammation around, reduce the slough to its smallest size and effect its rapid separation. If suppuration has occurred, it will lessen its area, disperse it, or quickly expel its products from the living body.

VIII.

TREATMENT OF IMMINENT INFLAMMATIONS AND SOME NON-INFLAMMATORY DISEASES.

IT is well known that the chief danger after certain injuries and operations is inflammation. After a wound of the abdomen, or, but in much less degree, the operation for ovariotomy, or the operations for strangulated hernia there is danger of peritonitis. This, I believe, I have sometimes prevented and sometimes diminished by the free application of iodine to the abdomen and cantharides to the femoral artery. In injuries to the joints, especially perforating wounds, which are frequently followed by destructive suppuration, does it not seem in the highest degree probable that a vigorous excitement of the tissue elements of the skin tends to lessen the excitement of the tissue elements of the wounded synovial membrane? In an injury of the head, is any one prepared to say that an artificial inflammation of the neck does not tend to divert pathological action from the intra-cranial contents?

There are a few non-inflammatory diseases in which derivation of blood by means of counter-irritation is of more or less service. In spermatorrhœa, which is referred, I think correctly, by Sir Henry Thompson, to congestion of the prostatic portion of the urethra, counter-irritation to the perineum and thighs I have found much more successful than any other treatment. In enlarged prostate and internal morbid growths of various kinds, a stripe of irritation over the next large artery will often relieve discomfort.

CASE I.

Delayed peritonitis after rupture of the intestine: Death from shock on the seventh day.—A woman of 35, after injury to the intestine, was insensible from profound shock for six hours, with an axillary temperature of 95°. When she rallied she progressed for several days very favourably, except that she was subject to occasional pain in the abdomen of a very severe character, which changed its locality at each recurrence. These pains were always relieved by vigorous local applications of iodine liniment around the seat of the pain and in the groins. Opium and a little nourishment were given by the rectum only. During the seventh day she suddenly and in a few hours sank from shock. After death a rupture of the intestine was found, and a thin layer of solid fæces spread over the intestines. There was no peritonitis. *Remarks.*—Here the irritated skin prevented peritonitis, and prolonged life several days.

Case II.

Ovariotomy during peritonitis: Counter-irritation at groins: Recovery.—Mrs. K., age 40, entered the hospital for the purpose of having ovariotomy performed. There was constant vomiting, a quickened pulse, and a temperature of $99\frac{2}{5}°$, with abdominal tenderness. As these symptoms did not abate with time, the operation was performed. Universal recent adhesions were found Vomiting and pain during the first few days were met chiefly by counter-irritation in the groins. Improvement uniformly followed each application. Linseed poultices to the abdomen were also sprinkled with mustard. Complete recovery followed. *Remarks.*— Here, as in previous case, it will be seen that there was no *post hoc propter hoc* delusion.

Case III.

Strangulated femoral hernia of several days' duration : Extreme inflammation of intestines : Counter-irritation over "femoral:" Recovery.—A female servant, age 40, was brought into hospital with very marked shock, and a strangulated femoral hernia of a week's duration. The intestine was of a dark blue colour, but being still firm it was returned. Acetum lyttæ was applied over the adjacent femoral artery. Recovery was uninterrupted. *Remarks* —This case is cited as fairly representing a class—a class in which there could be no reasonable doubt that counter-irritation over another vascular region was extremely beneficial.

Case IV.

An incised and gaping wound of the knee-joint: Rest, pressure, counter-irritation: Recovery.—A man, age 20, a currier, accidentally laid open his right knee anteriorly to 'he extent of two inches. The articular extremities were visible. The wound was carefully closed with silver wire, the joint was compressed, and a splint affixed. Counter-irritation to the thigh and leg was adopted and maintained. Not a drop of pus formed and no unfavourable symptom occurred. In six weeks he was allowed to get up. *Remarks.*—It is impossible to say that this case would not have done well without the counter-irritation. One circumstance however, struck the dressers and pupils. Counter-irritation was not applied for several hours after admission. He was complaining of pain in the limb, and was very restless; when the counter-irritation was excited, both the symptoms immediately disappeared and did not return. I brought this case before the Medico-Chirurgical Society, and as some remarks I made on injuries of the joints were misunderstood, I take this opportunity of correcting the misunderstanding. I was supposed, by a reporter at least, to say that the more severe the injury of a joint, the better. What I did say was that a given injury of a joint would do better if there were other injuries in the vicinity of the joint. A scarification of the skin of the thigh would be a counter-irritation. Many experienced surgeons have confirmed me in this view.

IX.

TREATMENT OF INFLAMMATORY DISEASES NOT PECULIAR TO ANY ANATOMICAL SYSTEM.

LOCAL INFLAMMATION.

A LOCAL inflammation which has not proceeded to suppuration or ulceration is not described in surgical text books, or any reference made to its treatment. I venture, however, to draw attention to inflammation in its earlier stages as a distinct disease. If we pass from the surgical library into the surgical ward, or out-patients' room, we frequently meet with enlargements which are tender, hard, hot, and perhaps painful, and covered by red or œdematous integuments, which often subside quickly or slowly, and which are more or less influenced by treatment. Such an inflammation is not an abscess, or a carbuncle, or an erysipelas, or, in the ordinary sense of the word, a cellulitis.

Often in such cases all that we can say is that an inflammation is present. But it is also not rare to find such a combination of local phencmena as enable us to do more than this, to say, for instance, here is an incipient abscess, here an incipient carbuncle, here an incipient whitlow, here an incipient cellulitis. There is yet no pus in the abscess or whitlow, no slough in the carbuncle or cellulitis.

In the great majority of these cases it lies in our power immediately to arrest the inflammation. We can not only prevent suppuration or sloughing, but we can, sometimes even in a few hours, rapidly remove considerable masses of induration and painful swelling. How can this be done? Chiefly by counter-irritation. A belt, or zone, or horse shoe of counter-irritation carried around, and a little distant from, the inflammation will remove it in a period of time which will depend on the extent and intensity of the new inflammation. If the inflammation be acute, a quickly acting irritant should be used, such as the acetum lyttæ. If the skin be moistened with this and allowed to dry, and then be moistened again, vesication will probably occur very rapidly. If in twelve or twenty hours the inflammatory symptoms have not distinctly subsided, a second or third application may be made. A little acetum lyttæ may be lightly painted over the first zone, or a new zone may be painted within or outside the first zone. The latter method is less painful. In less acute and chronic cases (the larger number) a wider zone or belt may be painted, first with iodine liniment and a pigment applied twice a day for the first few days. In children, the ordinary tincture of iodine, with or without a small quantity of the liniment, may be used. A single moderately free application of iodine paint (or acetum lyttæ to a some-what limited surface,) under the influence of chloro-form, will often prevent a large acute abscess in a child. In very young children the tincture of iodine is a sufficiently powerful application.

CASE V.

Extensive inflammation of popliteal region : Counter irritation to thigh and leg : Rapid subsidence.—A little girl of five was brought to the Queen's Hospital with the left knee semi-flexed. A large red, hard, and very tender swelling filled the popliteal space, and encroached on the thigh and leg. Smart fever was present. The thigh and leg were moistened (as in the diagram) leaving only the inflamed part uncovered, and a large poultice, when the paint had dried for an hour, was ordered to envelop the knee and adjacent limb, and to be surrounded by several turns of bandage, securing pressure short of pain. In two days nearly all inflammation gone ; in five days limb quite well.

F

Remarks.—Extensive counter-irritation was here instantaneous in its results. Probably in so large a tumour pus was present and was absorbed.

Case VI.

Severe incipient whitlow : Counter-irritation : Immediate subsidence.—A woman presented a finger in which a tense shining red and painful swelling had come on for a week. Iodine paint to the *other* fingers, all the hand, and part of fore-arm. Complete relief in twenty-four hours. All symptoms gone in a few days· *Remarks.*—The alternative treatment was a knife, poultices, and a three weeks' sling. This case is selected as representing a large number of similar inflammations, *all* progressing unfavourably to the moment of counter-irritation, then all immediately and rapidly subsiding.

Case VII.

Sub-acute glandular inflammation: Counter-irritation: Arrest and subsidence —A girl of four was ailing a few days, when a hard oval swelling was found occupying the outer two-thirds of the groin. A horse shoe of weak paint was applied above externally and below, but the skin was very sensitive, and only a little was used occasionally. A poultice was applied within the horse-shoe of paint. It caused smarting. No

marked change occurred for several days, when the
inflammatory mass rapidly disappeared, and the child
was suddenly well and active.   *Remarks.*—Such cases
mostly end in abscess of three weeks to three months'
duration.

Case VIII.

*Incipient abscess in axilla : Circumjacent counter-
irritation : Immediate disappearance.*—Jno. H., age 28,
labourer, had a large red, painful, and tender tumour
in the axilla, of fourteen days' duration.   Iodine paint
was freely applied to the thorax, axillary folds, and
upper arm.   On the third day the swelling had
disappeared, and on the seventh there was complete
recovery.   *Remarks.*—An example of a frequent
condition mostly ending in abscess of a very obstinate
character.   One of many in which suppuration has
been prevented.

X.

THE TREATMENT OF ABSCESS.

I SHALL here speak of abscesses generally, and such
as are mostly near the surface of the body.   Those
which are peculiar from their anatomical relations will
be considered separately.

It cannot be too well understood that abscesses present every grade of acuteness and chronicity, from the utmost acuteness to the utmost chronicity. I shall speak first of the more acute.

The treatment of the more acute abscesses is uniform and simple. If very acute, a not wide (the width must always depend on the extent of the primary inflammation), zone of integument should be painted with acetum lyttæ. A couple of moistenings, with a few minutes' interval, and a third in two or three hours, if vesication be not marked, will suffice. The benefit will perhaps show itself in twelve or eighteen hours, but then it will be very, indeed marvellously rapid. In somewhat less acute abscess (and even in the most acute) an effectual and rapid treatment is to paint a broader zone of iodine—the liniment, or a stronger pigment, and to repeat this once or twice daily as often as the patient can bear it. The directions I have just laid down are applicable to any abscess, but in certain localities there are a few modifications, or rather additions, that may be made with benefit. Thus, if an abscess be situated in the vicinity of a large artery (as in the limbs or neck), which large artery is more or less an independent, or "next" artery, a stripe of counter-irritation may with advantage be established over it. In an abscess (or carbuncle, or boil, or erysipelas) in the axilla, or in the thorax, near the axilla, a stripe of

counter-irritation—narrow of acetum lyttæ, broader of
iodine—may be carried along the brachial artery to the
elbow. In abscess about the elbow stripes of counter-
irritation may be established over the radial and ulnar
arteries. In abscess of the groin (bubo, strumous
glands, &c.), a stripe of skin may be inflamed over the
femoral artery. These stripes should be made in addition to the zones, or circles, which may in consequence
be made a little narrower. In abscesses of the hand or
fingers, all the uninflamed part of the hand and the
whole of the forearm (or half, or two-thirds, in small
abscesses), should be covered with iodine. At the same
time a linseed poultice thick, heavy, and hot, should
be applied to the abscess, and over it should be placed
a bandage, or a shot-mattress—which of the two con-
venience and locality may determine. In either case
the object is to secure pressure, which should be
moderately firm, but which also should never give rise
to pain. Real trouble should be taken to elevate the
part by light wood apparatus, or pillows, or both. In
abscess in the hand, or fingers, the hand should be
held up to the chin by a sling, if the patient be not
confined to bed.

With many surgeons the main question in the treatment of the more acute abscesses is—when shall they
be opened ? When the treatment now described is carefully carried out, the question is a very secondary and

unimportant one. *With this treatment, as a general rule, it is not necessary to open abscesses.* No advantage is gained by opening them. The ill results which, in the ordinary treatment of abscesss, are avoided by the knife —(indeed, the knife may be said to be the current, accepted treatment)—are best avoided by a circumscribing belt of counter-irritation assisted by the other remedies referred to. *A zone of cutaneous inflammation immediately removes the inflammation, and thereby immediately removes the pain, the pressure on adjacent parts, the tendency to either extension or diffusion, and the danger of opening into important cavities.* In a large proportion it is possible to obtain the absorption of pus. In others spontaneous opening quickly and readily occurs. In a small proportion, however, where, notwithstanding the removal of inflammation, there is no marked tendency to spontaneous opening, an incision may be made in the usual way. Uniform and moderately firm pressure is an admirable and rapid mode of opening an abscess.

The effect of counter-irritation on abscesses is very striking. It takes away the diffused swelling and hardness around the abscess, and leaves the pus in a circumscribed isolated cavity which projects from the surface with all the distinctness and prominence of a tumour. The skin is slightly red over the round swelling.

The illustrations represent a case where, in from thirty to forty hours after adjacent counter-irritation, a fore-arm more than double its natural size showed only a defined convex swelling, not a sixth of the previous size.   In such cases as this I have repeatedly seen the soft convex fluctuating prominence entirely disappear with a few days' continuance of counter-irritation, pressure and rest.

If there were no advantage in the rapidity of the cure of abscess by zones of counter-irritation, an unquestionable advantage may be claimed in the greatly diminished formation of pus.   When the knife is used the formation of pus is, in most cases, continued for at least several, often for many, days.   The swelling, hardness, and tenderness take several or many days to subside.   *When efficient counter-irritation is used, all the inflamed tissue which is not already pus is reclaimed from the suppurative process.*   Under such circumstances when an abscess opens, or is opened, scarcely any or no discharge follows the first escape.

I believe that incisions are often an excellent form of counter-irritation, and that it is in this way they are beneficial in the majority of abscesses.   In a small minority where abscesses are under dense and tense tissue, they have other and obvious benefits.   In the latter class of cases, if counter-irritation does not relieve pain and swelling in a few hours, by all means let them be resorted to. I am no advocate of timid surgery. But the counter-irritation of incisions is tardy in its effects—probably because the interval which intervenes between the incision and its resulting inflammation, is much longer than the interval—if indeed there be any—which intervenes between the application of cantharides, iodine, &c., and the inflammation which they excite. I have no doubt whatever that many an abscess has been cured by an incision which did not reach the suppurative cavity, and from which no pus at any time escaped.

If it can be affirmed that, as a rule, abscesses may be quickly and safely cured without the knife, it is no slight advantage to women, children, and timid men. Indeed I have often found medical and courageous men so depressed with a local painful inflammation, that they have insisted on the use of chloroform for the opening of an abscess. The treatment described is applicable to every stage of abscess, from the earliest to the latest, in which any inflammation remains. Patients often come to us with abscesses which have long been freely open, but where copious discharge, pain, tenderness, and swelling remain. While writing these lines I have seen a woman with a large abscess occupying the whole forearm, which has been freely open for a fortnight, with no improvement save a little less swelling.

In chronic abscess the principles of treatment are the same, but the counter-irritation should be milder, and the zone or other shaped surface larger. Iodine preparations are here of great service.

The uses of pressure, rest, elevation, and removal of cause, where practicable, should be carefully remembered. In chronic abscesses it may be more frequently desirable to make an opening, especially in strumous abscesses not dependent on disease of adjacent organs, bone, or other structures. Chronic abscess, which is very common, is that which accompanies osteitis and caries of bone. But the abscess here is only a symptom—indeed a trifling symptom—for if the caries could be removed the abscess would cause no trouble. In abscesses from caries of skull, or ribs, or sternum, or vertebræ, or tibia, it is the caries which determines our treatment.

The general treatment of abscess is that of inflammatory diseases—and has been already described—a nutritious, simple diet, stimulation slight, or not at all, except where very copious suppurative discharge and attendant exhaustion are present. When the condition referred to prevails, considerable use of stimulants may be required, a little brandy, *largely diluted with water*, is better than wines. In large abscesses rest in bed is very desirable, with fresh air if possible, and always a warm skin.

G

Case IX.

Acute abscess in infant: Circle of counter-irritation, opened spontaneously second day, closed the third.—Alice F. was brought to hospital with an abscess in gluteal region, the size of a hen's egg. A broad circle of mild iodine paint (tinc., four-fifths, lin., one-fifth) was applied around the abscess. The next day the abscess opened spontaneously, on the third day the abscess had quite disappeared, and the opening was closed. *Remarks.*—This case fairly illustrates a common result of the treatment described, namely, the rapid opening and closing of an abscess which seem to depend on the rapid removal of circumjacent inflammation.

Case X.

Large abscess of the palm: Counter-irritation to hand and forearm: Pressure and elevation: Very rapid recovery (four days).—Joseph G., age 7, came to hospital with a large abscess in the palm of the right hand. The hand, fingers, and forearm were swelled to twice their natural size. The abscess had been present sixteen days, and spontaneous opening had occurred ten days before admission. The hand, except the palm, and the forearm, were covered with iodine liniment. A linseed poultice, with bandage and sling, were also ordered. In forty-eight hours the discharge had ceased and the swelling had subsided. The hand could be used on the fourth day. *Remarks.*—The swelling and a copious discharge continued for ten days, after spontaneous opening. Immediate subsidence on the application of extensive counter-irritation was *not* a coincidence, it may be presumed.

Case XI.

Large abscess in an old man : Adjacent counter-irritation : Immediate spontaneous opening and recovery. —Amos S., age 92, presented himself at hospital with a large abscess at the side of the neck. The iodine liniment was applied freely to a comparatively large surface at the back of the neck. Two days afterwards the abscess opened spontaneously ; two days later still, that is, on the fourth day, the opening had completely closed and the swelling had disappeared. *Remarks.*— This case represents a large number of others, and the rapid recovery after counter-irritation was certainly no coincidence.

Case XII.

Strumous and chronic abscess of axilla : Recovery accelerated by counter-irritation.—Edwin O., aged 17, had a large strumous abscess of axilla, with extensive redness and induration around. Duration three weeks. A circumjacent zone of integument was painted with iodine liniment and a pigment prescribed for the patient's own use. Fourteen days later only a little induration could be detected. *Remarks.*—This case is important as illustrating a class in which the strumous diathesis is very marked, and where the accompanying abscess is extremely tedious and chronic—some redness, hardness, and discharge often continuing for months.

Case XIII.

Large abscess of face from carious tooth: Counter-irritation over carotid and brachial arteries with marked effect.—Hannah G., aged 35, married, had a large abscess over the jaw and neck, which was probably caused by a carious tooth, which however she refused to have removed. Counter-irritation was effected over the carotid and brachial arteries of the same side with immediate and marked effect on the discharge and swelling. *Remarks.*—This case teaches us that where even a mechanical cause is irremovable we can nevertheless often exert great control over inflammatory action.

Case XIV.

Two large abscesses of the forearm, with extreme circumjacent swelling, suddenly converted into two tumour-like prominences by means of counter-irritation.—James T., aged 35, a smith's striker, came to the hospital with two large abscesses on the forearm, both on the extensor surface, one near the elbow, the other near the wrist. Iodine liniment was applied to the forearm, except the more prominent portions of the abscesses, the hand and part of the upper arm. When he came three days afterwards, the arm presented a very singular appearance. The swelling of the forearm had entirely gone, and the abscesses were reduced to two circumscribed spherical prominences (see illustrations), over which the skin was slightly pink. They were then opened. In thirty hours the swelling and discharge had disappeared. *Remarks.*—These cases occurred nearly two years ago, before I clearly saw that the knife, in the great majority of cases, is unnecessary. The effects of counter-irritation in instantaneously removing all inflammation, leaving only the actual pus, were little less than magical.

Abscesses under Mr. Turton's care: Counter-irritation attended with great benefit.—Mr. Turton reports that "In two cases of large chronic abscess, which had become inflamed, the repeated application of iodine paint around has taken away all pain and procured absorption of a large portion of the contents."

Case XVII.

Abscess of hand: Counter-irritation to forearm, and abscess opened: Recovery in twenty-four hours.— Laurence M., aged 30, had large abscess on dorsum of left hand. The hand and forearm were covered with iodine liniment, and then the abscess was opened. After the first hour there was no discharge. In twenty-four hours the swelling had gone and the opening closed.

Case XVIII.

Large abscess in gluteal region: Circle of counter-irritation: Recovery in forty-eight hours.—William C., aged 18, had large abscess in gluteal region. A broad circle of iodine was applied. In twelve hours the abscess opened, and in forty-eight hours all the symptoms had gone.

PHAGEDÆNA, SLOUGHING PHAGEDÆNA, HOSPITAL GANGRENE, SLOUGHING, SENILE GANGRENE, GANGRENE.

The words at the head of this chapter are used with somewhat different meanings by different surgical writers. Perhaps there is something arbitrary in their use, but one thing is perfectly clear, and that is all that relates to the purpose of these pages. *They denote acute destructive inflammation,* whether the destruction be molecular, or of small, or large masses of tissue. They may and do occur in any part of the body, but they are a little more common in the vicinity of the genitals and in the lower limbs They may and do ensue on most varieties of inflammation, but it is a little more common for them to seize on those that are of a syphilitic character.

The syphilitic ulceration of the nose is usually a less rapid process than inflammatory disorganization elsewhere. A zone of acetum lyttæ, with an inner and outer (one or both) zone of iodine should be maintained until the process is completely checked. In the excessively rapid sloughing ulceration in the vicinity of the pelvis, as in the groin, or labium, or mons veneris, or nates, or penis (it is, however, most frequent in women), a broader zone, or couple of crescents, or horse-shoe should be painted with acetum lyttæ, and an outer and still broader zone, with iodine liniment. The iodine acts immediately, and the acetum lyttæ in a few hours. The principle of treatment in these cases should be this: the more acute the destruction, the more prompt and vigorous (acute) the circumjacent counter-irritation. A single broad zone of iodine liniment answers very well.

There can be no better application to the diseased
surface than a good linseed meal poultice. The less
the surface is teased with any kind of dressing the
better. *The true way of cleaning a phagedœnic or a
sloughing ulcer is to remove the inflammation,* then the
sloughs rapidly fall off and the liquids cease to flow.
Purulent or ichorrhœmic infection is never prevented
by clean dressing. It is effectively prevented by a
rapid arrest of the destroying inflammation. The free
use of antiseptic liquids at and around the inflamed
part is quite compatible with non-zealous dressing.

Is the application of strong nitric acid to a
phagedænic or sloughing surface of any use? It is
very much less useful than a vigorous zone of counter-
irritation, but in the absence of that it is an unquestionably beneficial treatment. It acts as a counter-
irritant, just as nitrate of silver does to the os uteri, or
the surface of an indolent ulcer, or an incision in
erysipelas. *The apparently direct irritation is in relation
to the mass of inflamed tissue a positive counter-irritation.*

But the ordinary "nitric acid treatment" has this
serious disadvantage—it produces itself a slough where
any further slough may be fatal. If sloughing action
be already close to the femoral artery, the artificial
slough produced by nitric acid may unquestionably
open it.

In exceptional cases sloughing may resist the treatment advocated here. In such cases the syphilitic diathesis is probably intense. Under such circumstances I have (in opposition to much book advice, but with excellent results) induced slight mercurial action as rapidly as possible by mercurial inunction, and small doses of mercury, the sixth of a grain of calomel every hour, for twelve, or twenty, or more hours.

Cancrum oris in its severer forms, and the analogous disease of the vulva in little girls is now happily rare. Should they present themselves, a circumscribing zone of active counter-irritation would, more rapidly than any other measure, check the inflammatory ravages. When senile or other gangrene presents evidence of unusual inflammation, adjacent counter-irritation may be tried.

Case XIX.

Sloughing Phagedœna of the leg : Counter-irritation : Immediate arrest of the destructive inflammation.—Maria T., aged 23, was attending the out-patient rooms with a cluster of deep tertiary ulcers on the outer side of the leg, a little below the knee, when the ulcers were attacked with phagedænic action. The ulcers rapidly coalesced, there was extreme pain, great surrounding swelling, copious discharge, and a dirty yellowish brown appearance on the surface of the ulcer. A ring

of blister was put round the leg above the ulcer. In
forty-eight hours the ulcer was clean, and the swelling, discharge, and pain had all gone. The subsequent
cicatrisation of the ulcer was much accelerated by the
use of iodine in a patch at the back of the leg.
Remarks.—This case came before me some time ago.
I should now use the acetum lyttæ in preference to
the ordinary blister, and apply it in the form of a zone.

Case XX.

*Severe sloughing ulcer of the leg : Counter-irritation
(probably too mild) of little use.*—Maria T., aged 24
(the same patient as in the last case, twelve months
later, and after an interval of health), came to hospital
with a large sloughing ulcer of the leg, ensuing on a
relapse of deep tertiary ulcers. The ulcer was on the
upper and outer part of the leg, and in size a little
larger than the top of an ordinary tumbler, portions of
the tibial periosteum and the ligamentum patellæ were
exposed. The base and margins of the ulcer showed
masses of dead tissue. The pain, discharge, and
swelling were extreme. A broad belt of iodine
liniment was painted around the ulcer, but the result
was not so satisfactory as my experience of other cases
led me to expect. There was no further sloughing,
but neither was there rapid subsidence of inflammation

nor cleaning. At the expiration of twenty-four hours I commenced giving small doses of mercury very frequently, continuing the iodine application. In two days the beneficial effects were very marked. *Remarks*. —In so destructive an inflammation I ought to have used acetum lyttæ freely and extensively. There was no doubt a strongly diathetic influence also to combat, for which I believe mercury to be the best remedy.

CASE XXI.

Phagedœnic chancre behind the corona, threatening the glans and urethra : Femoral blister, and immediate arrest of phagedœnic action. A man, aged 30, of cachectic appearance, came to hospital (he had not taken mercury) with a phagedænic ulcer, progressing deeply and rapidly, and close to the urethra behind the corona. Acetum lyttæ was applied over the femoral artery, for the extent of six inches by one. In twenty-four hours the destructive inflammation had disappeared, and the chancre pursued the course of an ordinary healing ulcer. *Remarks.*—The formation of an urinary fistula was certainly prevented in this case by counter-irritation, probably the vitality of the glans was also preserved.

CASE XXII.

Sloughing chancre: Counter-irritation: Immediate arrest.—Benjamin C., aged 26, came to hospital with extreme swelling and pain of the prepuce. Active destructive inflammation was present in the glans, which, prior to the attack of sloughing, was the seat of indurated chancre. Two longitudinal blisters were ordered over the femoral arteries, and in forty-eight hours all the active symptoms had disappeared. *Remarks* —In cases of this kind I now prefer the use of (glacial) acetum lyttæ; the action being more rapid, the results are accelerated in a precisely corresponding degree.

ERYSIPELAS, SIMPLE AND PHLEGMONOUS. CELLULITIS.

THERE are two points in the prevalent treatment of erysipelas which require a few words of comment. A favourite treatment with many surgeons, and one as successful at least, if not more so than any other, is that of the ring of nitrate of silver—the ring alone or the ring and a covering also. The method of treatment is a curious because unrecognised illustration of the value of counter-irritation. It is supposed that a ring of skin moistened with nitrate of silver offers a *physical* impediment to the spread of erysipelas. It would be more correct to regard the belt, narrow as it usually is, as a circumscribing counter-irritation —a counter-irritation having a beneficial influence on the erysipelas as it would have on an abscess or a sloughing ulcer. The broader the belt the more useful it is. The next point is the value of incisions in phlegmonous erysipelas. Although I refrain from making incisions in carbuncle and often in abscess, I have not yet relinquished their use in the tense brawny stage of phlegmonous erysipelas. Incisions here (as the incision in abscess) not only give exit to fluids but act as counter-irritants ; each incision is a counter-irritation in relation to the island of tissue around. Incisions may usually be rendered unnecesary if the circumvesting artificial irritation be sufficiently early, extensive, and vigorous.

The treatment of *simple erysipelas* is not to be too slightly estimated, because in the early stages it is impossible to say that, what seems a simple, may not become a phlegmonous erysipelas. Let, therefore, the application of nitrate of silver or iodine paint be be made with tolerable freedom. The nitrate of silver solution should be moderately strong (2 drs. to oz., with three or four drops of strong nitric acid). It should

be applied as a broad zone around the inflammation,
and may be carried lightly over it as well. If iodine
be used the liniment should be used as a belt, and a
milder pigment over the part. A circle of acetum
lyttæ, intersecting the iodine or nitrate of silver, and
keeping up and carrying on the artificial inflammation,
may also be adopted. In this and in the *phlegmonous*
variety there should be moderate pressure, with
bandage or shot-mattress over a thick linseed poultice
until active inflammation has passed away, and over
astringent lotions and cotton wool at a later period.
In the lower limbs pressure and elevation greatly
assist other measures. In the phlegmonous erysipelas,
should the inflammation proceed in spite of the
remedies indicated, or should the case when first seen
have reached the tense brawny stage, the counter-
irritation should be made more vigorous, and incisions
may be resorted to as already mentioned.

With respect to carrying counter-irritation over as
well as around inflammations, there is not much to be
said. Nitrate of silver answers well for such a purpose
because it is very superficial in its action ; a weak paint
or the tincture of iodine, may also be used. The really
efficient counter-irritation, however, is the circumjacent
zone.

The usual hygienic dietetic measures requisite in
severe inflammations are of importance, and if great

exhaustion should set in, stimulants with moderate freedom may be needed.

Cellulitis, although an inflammation of the debilitated is, in my opinion, a pathological condition distinct from that of erysipelas. Its treatment, however, locally and generally, is similar to that of phlegmonous erysipelas. It occurs more frequently in the upper extremities and upper parts generally, so far as I have seen, still the part affected should be elevated as efficiently as possible. Vigorous circumjacent counter-irritation, incisions possibly, pressure, gentle at first, firmer afterwards, poultices early, astringent dressings later, are the local measures. Depression in cellulitis is often extreme, for which brandy, with abundant water, is best. In this especially, and in all the severer inflammations, it may be desirable to ease pain and produce sleep by means of opiate preparations.

In erysipelas and cellulitis there is no drug which cures. The tincture of iron is a favourite preparation with many surgeons. I give iron in erysipelas because I give it in all, or almost all, inflammations.

CASE XXIII.

(Under Mr. Turton's care): Extensive cellulitis of fore-arm: Counter-irritation: Mr. Turton's remarks.— An old man (60), a collier, had severe cellulitis from

fingers to elbow, the swelling having a brawny character, and following a poisoned abrasion of the finger. The inflamed surface was painted with liquor ferri perchloridi. To the adjacent surfaces iodine paint was frequently applied. *Mr. Turton's Remarks.* —"I had only to make about three incisions of small extent. There was less sloughing, less pain, less constitutional disturbance, less subsequent depression, than I have ever before seen in such a case. When I say less, I mean that I do not think he suffered in any of the above respects more than one-fourth what he must have done under any other treatment."

Case XXIV.

Cellulitis of neck and occipital region: Treatment by counter-irritation: Favourable recovery.—Martha H. (aged 47), came into hospital with a large swelling occupying the back of the neck and the occipital region, and especially prominent on its right border. The swelling had been present three weeks, and was increasing in size. It was very hard, but uniformly so, and also extremely painful and tender. Iodine liniment was applied in the form of a horse shoe around the lower margin of the enlargement. The tendency to increase in size was immediately arrested, but there was no marked subsidence for seven days.

when the diminution in size was rapid, and in five days the patient was well. There was no suppuration or sloughing. *Remarks.*—The avoidance of suppuration or sloughing was very satisfactory. I have very frequently observed that if the effects of counter-irritation are delayed they are very rapid when they do commence.

CASE XXVI.

Carbuncular cellulitis of twelve months' duration cured by counter-irritation — George C., aged 52, great uniform enlargement of calf of leg, which is very hard, the skin is white, but shining. Tenderness and pain not marked. Has been present and increasing for more than a year. Three weeks' rest, and counter-irritation to other parts of the leg, were followed by complete subsidence. During treatment, a soft spot in the centre was punctured with a grooved needle. Nothing escaped at the time, but in a few days a small slough escaped, with scarcely a drop of pus, and the opening immediately closed. *Remarks.*—This case is interesting pathologically, as it shows that there are local inflammations which cannot be correctly placed in any existing classification.

CASE XXVII.

Cutaneous Erysipelas : Counter-irritation : Immediate subsidence.—Mr. N., aged 42, an engineer, while

I

directing railway works in India, sustained in an accident a simple fracture of the right femur and a compound fracture of the left tibia. On his return to England, some months after, I found both fractures firmly united and in good position. A sinus remained over the seat of the tibial fracture. This was almost healed with rest and elevation when the patient determined to use his leg, and a smart attack of erysipelas, from ankle to knee, was the result. A broad band around the knee and another around the foot were painted with acetum lyttæ, and the erysipelaous surface itself painted with a solution of nitrate of silver. In twenty-four hours the disease had entirely disappeared.

Case XXVIII.

Erysipelas of face and head with œdema of the glottis and threatened suffocation: Counter-irritation and relief in twenty minutes.—I was called to make an opening in the windpipe of a man, aged 50, at the Queen's Hospital, who was in great danger of death from asphyxia, and who, at the same time, had severe erysipelas of the head and face. Before opening the air tube, I determined to use smart counter-irritation (iodine liniment) over the carotids, the upper part of the thorax, and the brachial arteries, and if there was no benefit in fifteen minutes, to operate. In ten minutes the breathing was markedly easier, and in twenty minutes was comparatively calm. The patient subsequently sank from exhaustion, and after death extensive renal disease was found. *Remarks.*—In adults erysipelas is so often associated with visceral disease that treatment of any kind must often fail. Here the rapid effect of counter-irritation was very striking.

SYPHILIS.

In constitutional syphilis, from the indurated chancre onwards, the specific element so greatly predominates over the inflammatory as to require the treatment to be necessarily of a specific character.

After careful attention, directed to the subject for many years, I do not hesitate to say that the *mercurial* is the only satisfactory treatment of constitutional syphilis.

But there are not unfrequently syphilitic local conditions in which inflammation, as such, largely enters. In the suppurating chancre, a stripe of iodine paint over the femorals will often remarkably curtail the ordinary duration of the disease. In a case of

large, freely suppurating chancre, the size of a shilling,
on the body of the penis, in which I applied iodine
paint over the femoral artery (a stripe six inches
by two), the discharge entirely ceased in three days.
The sore was covered by a scab, which was attended
by neither pain nor tenderness, and which fell off
in six days. In the suppurating bubo, a horseshoe
of iodine (liniment or strong pigment), with rest
and uniform pressure, by means of a linseed
poultice and shot-mattress, will often effect complete subsidence in an early stage, and will rapidly
open and heal a bubo at a later stage. Uniform and
moderately firm pressure is an excellent and quick
means of opening an abscess where opening is
unavoidable. I have seen now several cases of
absorbed abscess and bubo, and only good results,
locally and generally, have followed. Secondary and
tertiary ulcers, in addition to constitutional treatment,
may be benefited by zones of counter-irritation, and
the more active the inflammation the greater the
benefit. The treatment of destructive ulcerations,
syphilitic as well as simple, has been already
described.

Case XXIX.

*Syphilitic periostitis of nose immediately relieved by
counter-irritation.*—Mr. C., aged 25, single, while
suffering from secondary skin eruptions, especially on

the forehead and the back of the neck (round the hairy scalp), was seized with sudden and severe pain and tenderness in the locality of the nasal bones, accompanied by very marked œdema. A very vigorous patch of counter-irritation was at once established at the back of the neck by means of the iodine liniment. In six hours the pain ceased, and in twenty-four hours the tenderness and œdema had nearly disappeared.

CASE XXX.

Suppurating bubo (with secondary syphilis) rapidly absorbed by means of counter-irritation and pressure.— A gentleman, aged 25, came to me with secondary syphilis, and a large fluctuating swelling on the left groin. I had never seen so large a swelling in the groin. With the circumjacent œdema it measured nine inches in diameter, and encroached considerably on the thigh and abdomen. A broad horse shoe was painted round it, and the irritation carefully kept up. A poultice with a shot bag was applied, and perfect rest maintained. In four days the swelling was only three inches in diameter and singularly circumscribed (tumour-like) with a pink colour of skin. In another week the swelling entirely disappeared, leaving, in the centre, loss of sensation, for a time.

Case XXXI.

Suppurating bubo : Counter-irritation over the femoral artery : Very favourable result.—George D., age 20, came to hospital with bubo discharging freely. A longitudinal blister was applied over the femoral artery. In three days the suppuration had entirely ceased. The subsequent healing was much more rapid and favourable than is commonly seen.

Case XXXII.

Persistently tender cicatrices after deep tertiary ulcers : Circumjacent counter-irritation : Immediate relief.—Elizabeth Y., age 28, married. After the healing of a cluster of deep tertiary ulcers near the knee in both legs, there remained very obstinate redness and tenderness. A belt of iodine irritation was established and maintained a few days, with complete relief. *Remarks.*—The tenderness was distinctly inflammatory, not hysterical, hence the rapidity and completeness of the relief.

Case XXXIII.

Syphilitic laryngitis and ulceration of fauces : Counter-irritation : Rapid recovery.—Joseph P., age 30, married, applied at hospital with great swelling and ulceration of fauces, associated with great difficulty of breathing and hoarseness. A blister was applied to the angle of each jaw. In forty-eight hours the relief to all the symptoms was very marked. In fourteen days, with a mild mercurial course and counter-irritation, the case was well.

CASE XXXIV.

Soft chancres and gonorrhœa : Counter-irritation : Rapid recovery from both.—Thomas J., age 24, single, came to hospital with two freely suppurating soft chancres, and a copious urethral discharge, with severe scalding. A blister was placed over each femoral artery for the purpose of removing the gonorrhæa, and with no anticipation of affecting the chancres. The scalding and most of the discharge disappeared in forty-eight hours. The discharge of the chancres dried in the form of a crust; they and the gonorrhœa were quite well in a week. *Remarks.*—The effect of the counter-irritation on the sores was striking and unexpected. They were of recent origin.

CASE XXXV.

Syphilitic ulceration of the throat : Counter-irritation to the angles of the jaws with great benefit.—Charles T., age 40, widower, had syphilitic ulceration of the throat, with more than usual swelling and discomfort.

Iodine was applied to the angles of the jaws, with immediate subsidence of all active inflammation.

Case XXXVI.

Indurated chancres, gonorrhœa also: Counter-irrita-tion: Great benefit.—Daniel F., age 24, single, had three chancres which formed successively, and which were accompanied by well defined induration. A copious discharge and scalding were also present. The inguinal glands were enlarged and painful. Iodine paint was applied to the groins and along the femorals. The discharge and the scalding completely ceased in forty-eight hours, as well as the pain and tenderness of the inguinal glands. Curiously enough, one of the chancres suppurated freely at a later period. *Remarks* —Here the gonorrhœa and glandular pain were instantly cured by counter-irritation. The case is otherwise of great interest in relation to the pathology of syphilis.

Case XXXVII.

Bubo of gonorrhœa: Counter-irritation: Entire sub-sidence without opening.—Robert L., aged 58, widower, had a large bubo, with unmistakably fluid contents. A broad horseshoe of iodine paint was applied around it, and directions given for a daily application, so as to maintain a moderate degree of soreness. Four days afterwards the swelling had almost disappeared.

CASE XXXVIII.

*Double bubo from soft chancre: Counter-irritation:
Rapid progress.*—William S., aged 19, single, had
several soft chancres and two large buboes of several
weeks' duration, and attended with considerable
circumjacent induration, and some, but not abundant,
discharge. The changes in both were of a very
chronic character. Both were treated with broad
iodine horseshoes, and in fourteen days both were
well.

CASE XXXIX.

*Syphilitic ulceration of nasal septum arrested by
counter-irritation.*—Emily S., aged 24, married, had
destructive but not very rapid ulceration of nose, the
septum being partially destroyed. Great and immediate relief was obtained by the application of iodine
liniment at the back of the neck. The ulcer did not
heal rapidly, but the destructive inflammatory action
was removed. *Remarks.*—This case illustrates well
the principle I have frequently drawn attention to,
the *inflammatory* element is readily removed by the
method here advocated, the *specific* element not so
readily.

CASE XL.

*Suppurating bubo with indurated chancre: Counter-
irritation: Rapid opening and closure.*—James M., aged

K

20, single, applied with a bubo in process of suppuration, an indurated chancre being also present on the prepuce. In addition to the constitutional treatment, a broad belt of iodine was painted around the bubo. The immediate effect was rapid pointing and opening, with equally rapid subsidence of the discharge and closure of the opening. *Remarks.*—A good illustration of the rapidly favourable progress of abscesses when inflammatory action is arrested.

CASE XLI.

Acute syphilitic destructive inflammation of fauces quickly relieved by counter-irritation.—David A., aged 25, single, had syphilitic ulceration of the fauces, attended with great swelling and severe pain, on swallowing or speaking. The left side was much the worse. A small blister was placed over the angle of the jaw on the left side. As the blister rose, the swelling, difficult swallowing, and difficult speech all rapidly disappeared.

CASE XLII.

Under Mr. Turton's care: Bubo with " evident fluctuation" absorbed by means of counter-irritation.—Mr. Turton writes : " A man came to me with a fluctuating swelling of the size of a small hen's egg, red and very painful. I applied the paint freely from the crest of the ilium down the front of the thigh. In two days he came saying that the dressing had taken away the pain. On examination I found the swelling much diminished, and repeating the application every second or third day for a fortnight the tumour was reduced to less than the size of an almond." *Remarks by Mr. Turton.*—"In this case there was fluctuation at first, and I merely expected to hasten suppuration by the strong application I made in the first instance. I have treated many of these cases, of which the above is a fair example."

X.

THE TREATMENT OF INFLAMMATIONS OF THE CUTANEOUS STRUCTURES.

CARBUNCLE.

CARBUNCLE may be to a remarkable degree controlled by the mode of treatment set forth in these pages, whatever the size, or intensity of inflammation, or stage of progress.

At whatever stage, then, a carbuncle presents itself, the one chief great remedy is a zone of counter-irritation. A circle of iodine liniment or strong iodine paint, several times repeated, will greatly relieve all the symptoms in twenty-four hours. The pain is instantaneously removed, all discomfort disappears in two or three days, and in a very few days longer the carbuncle is practically well. If the treatment be adopted when the first hard red swelling appears, the hardness, redness, swelling, pain, tenderness, all subside in a few hours. The wider and smarter the zone of counter-irritation the more rapid the subsidence. A strong solution of nitrate of silver would probably answer the same purpose, or a circular blister, but this is tardier in its action. If the carbuncle has proceeded so far that more or less cellular tissue is already dead, the slough rapidly separates from the living tissue, which is suddenly deprived of its inflammatory action by counter-irritation. The appearance of a carbuncle at this moment is very striking; the yellow slough in a few hours is protruded in a comparatively dry state, the ordinarily copious liquids which indicate severe inflammation being conspicuously absent. A healthy granulating surface only remains. Where the carbuncle is very large, the inflammation at its height, and the sloughing has just commenced, the treatment I adopt is no less remarkable in its efficiency and rapidity, although its duration is naturally a little longer than is required under more favourable circumstances. This simple treatment, a belt of iodine, will cure carbuncle in a fourth, or a sixth, or an eighth of the time required by ordinary measures.

In addition, a thick linseed meal poultice should be applied, large enough to cover the site of actual inflammation, but within the zone of iodine, as the water of the poultice disadvantageously dilutes the artificial irritant. Over the poultice a bandage should be applied, and for several hours in the day a shot mattress, as heavy as can be tolerated, should be placed over the poultice. Rest, at least of the site affected, is desirable, and as much elevation as is convenient. Happily carbuncles occur mostly at the upper parts of the body, and thus are naturally elevated. Scarcely any dressing is required ; the counter-irritation separates the slough better than scissors, and dries up the liquids better than a sponge. A weak disinfectant application occasionally, if there be time and need for it, will do no harm.

A simple nutritious diet, without stimulants as a rule, warm clothing, and, in severe cases, mental and bodily repose, are the chief items of the general treat-ment. Opium can rarely be needed, because the counter-irritation effectively relieves the pain. If counter-irritation be used in the evening it should not be too near bedtime, so that if there be any attendant smarting it may subside before the hour of sleep.

I have never incised a carbuncle, and I have never seen any benefit from the measure.

BOILS.

A LARGE boil is as severe and inconvenient as a small carbuncle, and numerous boils often give rise to great depression and discomfort. The treatment should be conducted on a similar principle to that of carbuncle, and indeed of all inflammations. A comparatively narrow zone of iodine liniment, a little wider one of a strong pigment, or a still wider one of a weak pigment, is the only local, perhaps the only treatment required. The zone will also vary in width in proportion to the extent and severity of the furunculoid inflammation. In a case cited below, a boil on the back of the forearm was accompanied by a swelling which involved two-thirds of its surface. In a case like this the iodine is put on the whole of the dorsum of the forearm, except an inch or two of the apex of the induration, and also over a portion of the flexor surface. When there are several scattered boils, a zone of counter-irritation is placed around each. When they occur in a cluster the iodine may be painted between them, and a wider zone established around the whole number.

In the treatment of *malignant pustule* I have no experience of the ultility of counter-irritation and pressure. This comparatively rare disease is so formidable and so uniformly fatal from its constitutional effects, that I should not be sanguine of the results of any possible treatment. Judging from the effects of the treatment, here advocated, in severe cellulitis, carbuncle, and phagedæna, I can say with confidence at least this, it certainly offers a better chance of success than any other. If a case came before me now I should, as the disease is a desperate one, adopt extremely vigorous measures. I should surround the pustule with a *broad circle* of acetum lyttæ, and outside that a broader belt of pure iodine liniment, using these remedies under the influence of chloroform if desirable.

CASE XLIII.

Large boil on forearm : Counter-irritation to forearm : Rapid recovery.—Lucy C., aged 46, came to hospital with a large boil on the dorsum of the forearm, in an early incipient stage. The remainder of the forearm, both extensor and flexor surfaces, were covered freely with iodine liniment, and directions were given for the frequent application of iodine paint. *In three days* all signs of active inflammation (swelling, hardness, pain, and tenderness), had disappeared—no slough and no pus having formed — leaving only a small dried epidermic crust in the centre.

CASE XLIV.

Carbuncle on neck : Counter-irritation : Recovery in seven days.—Eliza W., aged 48, came to hospital with

large carbuncle on the back of the neck. Sloughing
had not actually occurred. A band of iodine paint
was placed around it, and in three days suppuration
had taken place in the centre, and an opening had
formed. The band was made broader, and in four
days there remained only a small dry scab, all the
swelling, induration, pain, tenderness, and discharge
having subsided. Constitutional syphilis was present.
Remarks.—This case shows well the conversion of the
sloughing into the more favourable process of suppuration.

CASE XLV.

*Large boil on nates : Iodine zone four inches wide :
Rapid recovery.*—Joseph B., age 33, while attending
hospital with constitutional syphilis, a very large boil
formed on the nates four inches in diameter. A zone
of iodine liniment was painted around it, the zone was
of irregular width, broader towards the trochanter and
thigh, narrower towards the cleft and perineum. In
the course of five days central suppuration, opening,
closing, and subsidence of all symptoms occurred.

CASE XLVI.

Carbuncles under Mr. Turton's care.—Mr. Turton
writes thus : " I call to mind three cases of carbuncle
at the back of the neck which got well *without
incision or suppuration*,* by the liberal use of the paint

* The italics are mine.—F. J.

around the part and down between the shoulders.
Relief of pain is one of the most uniform results of
the treatment. Patients with carbuncle have always
come for the repetition of the application, saying how
much easier they were after it was done."

Case XLVII.

*Very large boil: A broad belt of iodine: Cured in
four days.*—William P., aged 19, came to hospital
with a very large boil on the left forearm—its extensor
surface. The swelling was three inches in diameter.
The induration, pain, and tenderness were marked.
It had been present eight days. The whole of the
forearm, except the boil, was covered freely with
iodine paint. In twenty-four hours the centre
suppurated and burst, the discharge ceased entirely in
forty-eight hours. In four days the boil was well,
only a small dry crust covering the centre. *Remarks.*
—This certainly remarkable case was not a " coincidence," simply because such a boil as this never did
get well before in so short a time with or without any
other treatment.

Case XLVIII.

*Large carbuncle on the neck: Counter-irritation:
Very rapid recovery.*—Ellen S., aged 37, married,
came to hospital with a large carbuncle on the right

L

side of the neck.   The hardness, pain, tenderness,
and almost purple skin, were striking.   A broad horseshoe of iodine liniment was freely applied in front,
below, and behind the swelling.   In consequence of
some misunderstanding, she did not repeat the application.   After the single application, however, the pain
immediately ceased, and the tenderness and discharge
and swelling rapidly diminished.   When she came on
the fourth day, a small yellow slough, without discharge
or swelling, projected through an opening in the skin.
A second application of iodine was made, six hours
after which the slough fell out, leaving a healthy
ulcer, which quickly healed.   *Remarks.*—No case,
having reached the same stage, ever got well so
quickly before.

Case XLIX.

*Carbuncle : Circle of counter-irritation : Rapid
recovery.*—George P., aged 55, had a large carbuncle,
of fourteen days' duration, on right forearm, with no
tendency to separation of slough.   The application of
iodine to the forearm and hand was followed, in three
days, by clean separation of the slough.   In six days,
only a simple healing ulcer was left.

Case L.

*Large boil on forehead: Counter-irritation : Rapid
recovery.*—Francis E., aged 12, had a large boil on

the forehead, with great œdema of the eyelids. A large half circle of iodine was applied. In three days, the boil and œdema had disappeared.

ONYCHIA.

THIS disease, in its several varieties, may be very advantageously treated with counter-irritation. The application of iodine—the strength varying with the age and severity of the disease—to the finger and hand, will very greatly curtail the ordinary duration of the disease. To the swollen end of the finger itself, pressure, gentle enough not to produce pain, is the best local treatment; the pressure may be made with a narrow bandage, or with strapping—a little lint, greased with mercurial ointment, being first applied to the ulcerated surface and its vicinity.

Onychia is of a syphilitic character in the adult, and in the young is very probably due to hereditary syphilis (whether it ever occurs without any syphilitic taint is very doubtful), so that a little mercury is of the greatest service.

PARONYCHIA.

EVERY variety of paronychia should be treated on the same principle. The only difference in the detail is the degree of vigour which should characterise the treatment, and this must be determined simply by the degree of intensity of the inflammation which is present. The thecal variety is, it need not be said, that in which the symptoms are most urgent and in which the phalanges and articulations are most likely to be involved. Suppose, then, such a case presents itself: the whole finger is swelled, hard, red, and intensely painful, the other fingers and the whole of the hand (dorsal and palmar surfaces) should be freely painted with iodine liniment, or a strong solution of nitrate of silver. The affected finger may be painted or not. A thick linseed poultice should be applied with moderate pressure. The hand may also be enveloped in cotton-wool, and somewhat compressed. The hand should then be carried under the chin and supported there with a sling.

In a still more severe case, where the hand also, as well as the finger, is swelled to perhaps twice or thrice its natural size, with suppuration probably present, *the forearm, almost to the elbow*, and the least inflamed portions of the hand, should be freely covered with iodine liniment, followed by poultice, pressure, and the chin-sling. Immediate subsidence of pain and of much of the swelling occur; if suppuration have occurred beyond a certain degree, the abscess will probably open within twenty-four hours of the iodine application, and, having opened, will immediately close without any continuance of discharge.

With the treatment for whitlow, just described, incisions are often unnecessary. The objects of incisions are two, to relieve pain and to save the deeper structures, bone, or joint, or tendon; both these objects are gained more rapidly and more pleasantly by a quickly acting counter-irritation. In a few exceptional cases, where subsidence of pain and swelling are not immediate, an incision should be made without delay. In these cases, the subsequent recovery is greatly accelerated by the counter-irritation. As paronychia is a good example of the more acute varieties of inflammation, and associated with little of the specific or diathetic element, so also it furnishes an excellent illustration of the benefit of the treatment I desire to enforce. Sometimes a single application of iodine suffices to remove all the symptoms; more frequently several applications are required.

CASE LI.

Severe whitlow of thumb : Counter-irritation : Subsidence in twenty-four hours : Relapse and ordinary treatment but no progress : Counter-irritation again, and again immediate subsidence.—A young woman, a factory worker, was under the care of an observant house-pupil (Mr. E. Smith) with severe whitlow of the thumb. He treated it according to the method advocated in these pages, with entire subsidence in twenty-four hours. Resuming her occupation too quickly she had a relapse, the ordinary treatment was adopted. After several days, without any changes, she said to Mr. Smith, " You will do no good, sir, without some more of that paint." With the "paint" immediate and permanent recovery followed.

CASE LII.

Severe whitlow: Counter-irritation to hand and forearm: Rapid recovery in four days.—Edward P., aged 64, came to hospital with hand and forearm twice their natural size, and a large abscess in the ball of the thumb. All the hand, fingers included, and forearm were covered with iodine liniment, moderate pressure, linseed poultice, and elevation were also resorted to. A few hours after, the abscess opened spontaneously. On the fourth day, the thumb was well, all swelling had subsided, and the discharge had ceased almost immediately after the abscess had opened. *Remarks.*—Is it too much to say that this case was cured in one-sixth of the ordinary time ?

CASE LIII.

Whitlow of thumb: Counter-irritation to hand and forearm: Recovery in forty-eight hours.—Hannah T., aged 25, had a severe whitlow of the thumb. After two applications of iodine paint to the hand and forearm, the symptoms subsided within forty-eight hours.

Case LIV.

Severe whitlow of thumb : Counter-irritation to hand and forearm : Rapid recovery —Sarah Ann L., aged 14, had severe whitlow of thumb with swelling of hand and forearm. Iodine liniment was applied to the hand and forearm. On the second day, an opening was made, and on the third day, the swelling and discharge had entirely disappeared.

ULCERS.

I OMIT any special reference to ulcers because, although inflammations, they are so much dependent on specific or mechanical and anatomical peculiarities that they mostly require specific treatment, and treatment especially directed to the removal of the cause. If inflammatory action should accidentally predominate, circles of counter-irritation and perfect rest and elevation would certainly do more to remove it than any other treatment. The treatment of destructive inflammations will be found elsewhere.

XI.

The Treatment of Inflammations of the Bones.

Perhaps in no department of surgery is there more difference of opinion, or greater diversity in the meaning and use of words, than in that of diseases of the bones. A few remarks are therefore necessary here, that my readers may clearly understand the pathological conditions to which I refer when I use the current phrases. A few surgeons apply the term "acute ostitis," and I believe a larger number prefer the phrase "acute periostitis," to the following condition : An acute inflammation attacks the shafts of the long bones, one or more, chiefly in the young, chiefly in boys. There is profuse suppuration, separating the periosteum, and occurring also in the intermuscular spaces, the shaft, all or part, rapidly dies. Whether the disease begins in the periosteum or the bone, or in both, cannot be discussed here, but it will be denoted in these pages by acute periostitis. Inflammation of the bone, ostitis, is a process which has several stages and results, which are slightly modified by locality, which merge into each other, or at least are not separated by sharp lines of demarcation, and which do not differ essentially from the phenomena of

inflammation in the soft parts. Inflammation of bone
—*ostitis*—refers to any state varying from increased
cell action, with its attendant vascularity, to commencing local death. This condition may subside, or
an increased quantity of bone—condensation—*sclerosis*
—may result. If the ostitis should proceed · to local
death, the characters of the dying and dead bone will
be modified by the structure of the affected bone.
Cancellous tissue, from its greater vascularity and
vitality, dies slowly and in minute fragments—*caries*.
Compact tissue dies rapidly, and in large masses—
necrosis. The difference is one of degree, not of kind.
Necrosis is sometimes found also in cancellous tissue,
and caries and ulceration (superficial caries) in compact
bone. A circumscribed ostitis leading to caries and
ending in circumscribed suppuration, and occurring
mostly in the head of the tibia, is called *abscess*.
These " phases" of bone inflammation may not only
merge into each other, but they often appear in conjunction. A mass of dead bone may be in a suppurating cavity ; around this is caries, externally to this,
condensation, and, still more externally, vascularity.

It is scarcely necessary to remark that certain of
the conditions just described can be treated by operative
means only, as complete necrosis, advanced caries,
fully formed abscess. Other conditions, however, and
in actual practice these are far more numerous, very

M

frequently admit of removal by non-operative remedies. I firmly believe that remedies, based on the principles laid down in these pages, are more frequently, more certainly, and more rapidly successful than any hitherto in use. We are urgently called upon to attempt to subdue any inflammation which has not yet given rise to irretrievable local death, as in acute periostitis or acute ostitis, in their earliest stages, although the prospect may not be very encouraging. But the prospect is really very encouraging in a great number of cases of ostitis (especially of the extremities of bones), in cases of commencing caries and actual caries in the young, and in cases of the less acute forms of periostitis. It is not amiss to note here that so-called hypertrophies of bone are almost invariably inflammatory diseases.

ACUTE PERIOSTITIS.

IN the acute inflammation of bone shafts, which is rapidly followed by suppuration and necrosis, it is only in the earliest stages that treatment can be of service. To be of service at all it must be vigorous and proportionate to the violence of the disease. Should a case in its earliest stage come before me, I should apply the iodine liniment very extensively, frequently, and very freely to the *whole limb*. Chloroform would be desirable, the more so the younger the patient

In these cases, incisions can easily be added to the treatment, and they should be made without delay whenever fluctuation can be detected.

Chronic Periostitis.

In all the less acute varieties of periostitis, great good is to be expected from treatment. The variety of chronic periostitis most commonly met with is the syphilitic—the node. The nocturnal pain of node is a most distressing symptom in a large proportion of cases. The iodide of potassium is an excellent remedy in such cases, but in my experience it is not a rapid one. I have often seen it fail, even as a constitutional remedy, when mercury has succeeded. Over and over again I have seen cases in which the pain has been uninfluenced by iodide of potassium, but *I have never known a broad belt of iodine liniment to fail to reduce pain within a few minutes of its use.* I have several times known a simple application *permanently* relieve the pain. Under the use of an iodine zone the swelling and tenderness also rapidly subside and the node disappears.

Whatever the state of the node may be, slight or severe, early or late, the zone of iodine is incomparably the best treatment. Should suppuration have occurred when the disease first presents itself, circumjacent counter-irritation is still the chief remedy, and incisions should be carefully avoided. If some exfoliation is taking place, the process of separation, although not a quick process, will be rendered safer and quicker by the same measure   Rest and pressure locally, and iodide of potassium—in large doses if necessary, and if these fail, mild mercurial doses with a non-stimulating diet are also remedial influences of great importance.

Ostitis.

I SPEAK here of an inflammation of bone mostly of a chronic or sub-acute form, chiefly affecting the extremities of the long bones and the short bones, and which, if it is not subdued early, proceeds to sclerosis, or, which is more important, to caries or abscess.   I need scarcely say that every surgeon sees a large number of cases of ostitis of the great trochanter, or lower third of the femur, or upper portion of the tibia, or the tarsal or carpal bones, or of the vertebræ or ilium—cases in which there is unquestionable inflammation (enlargement, pain, tenderness, impaired use), but in which there certainly is no caries.   In the extremities of the long bones the adjacent joint is often affected, nearly always in children, but very far from always in adults.   Such cases, too, if neglected, and in rare cases in spite of all care, pass on to caries or abscess.   The fact remains, however, that a large number of cases of simple ostitis present themselves for treatment. Without any doubt almost all of them can be cured. The presence of sinuses is not a certain indication that ostitis has merged into caries. The treatment for ostitis will often remove them.

At present rest is almost the only remedy : it must certainly be the foundation of all treatment. It is simply, however, a negative agency—it gives nature fair-play. But we can do something in these cases which nature cannot do. In counter-irritation we have a positive remedy. Iodine liniment is a convenient application in these cases. Giving rise to smart counter-irritation, it need only be applied every three or four days, and consequently requires less disturbance of whatever splintage is used. Splintage of gutta percha, or starch, should "fix" the adjacent joints —it should be light and permit of easy removal or readjustment. The iodine should be applied over a large surface. I happen to be speaking now of a disease in which iodine is occasionally used at present, but it is used over the inflamed bone only, where it should be least freely applied. If the great trochanter be inflamed, an oval belt should be applied, stretching upwards to or above the crest of the ilium, downwards to the middle of the thigh, forwards to the pubic spine, and backwards to the fold of the nates. The iodine may be applied over the trochanter also, but a little less vigorously. If the lower end of the femur or upper extremity of the tibia be affected, the lower half of the thigh and the upper half of the leg should be covered *all round*; if the carpus be affected, all the hand and half or two-thirds of the forearm. In the spine, a long vertical oval patch should be painted. In the spine, however, where the disease is deep, and where undisturbed rest is most urgent, a couple of long stripes of the actual cautery on each side of the inflammation is perhaps preferable. Indeed, in many cases the actual cautery may be substituted for iodine, (and always should this fail), especially in hospital practice, and of course never without chloroform. The mode and extent of applying iodine to any inflamed bone may be inferred from what has already been said.

CARIES.

CARIES of bone, however unquestionable its existence, should only be subjected to operative measures when other means have failed. Other means will frequently succeed. Caries in its early stages may often be cured. In the young, a case of caries which refuses to be cured must be looked upon as exceptional.

The treatment is so entirely similar to that just described, under the head of ostitis, that what was said there need not be repeated. Where ostitis has

advanced to caries, the process of cure will of course
be longer and the result not so certain, but efficient
counter-irritation with complete rest offers a prospect
of success which no other treatment, and certainly not
rest alone, can promise. The presence of sinuses
should not prevent the treatment being carried out.
It is commonly taught that when sinuses are present
it is too·late to resort to counter-irritation. Apart
from the fact that sinuses often occur in ostitis prior
to caries, I believe this teaching has done great
mischief. Granting that counter-irritation occasionally
fails (as it will if very limited, as it mostly did before
the value of rest was known), why should we
relinquish a powerful means of subduing inflammation,
unless we are prepared to relinquish all treatment
except operative measures? If the state of the caries
is not such as to justify surgical interference, why
should we avoid counter-irritation more than we
should avoid rest? It is true that counter-irritation,
where there are sinuses, often fails, but it is equally
true that rest also often fails. Is it logical that, because
the disease is more severe, the treatment shall be less
vigorous? It is commonly taught that when abscess
has formed in spinal caries it is too late for counter-
irritation. To me this proposition is simply astounding.
Should we not rather, seeing that the case is desperate,
and an operation out of the question, apply the actual

cautery to thrice or ten times the extent we should apply it prior to the appearance of a lumbar or psoas abscess?

An *abscess* in the bone implies a degree of inflammation still more advanced than caries, and, as a rule, requires a surgical operation. If only ostitis or caries be found by the trephine, the operation is admitted to be not only justifiable but beneficial. Where the abscess is small, the symptoms severe, and trephining is resorted to without finding pus, I should gouge out a considerable portion of the cancellous tissue, because I believe this to be a perfectly safe proceeding, and because the alternative, where there is intense and persistent pain, is loss of the limb.

Unless the severity of the symptoms call for early operation, the application of the actual cautery (under chloroform) in a cross, or circle, or patch over the painful part might quite possibly relieve the symptoms and cure the disease.

In ostitis, caries, and abscess, the iodide of potassium with iron, or the bi-chloride of mercury with iron, may be advantageously added to a nutritive and non-stimulating diet.

CASE LV.

Ostitis of fibula, its lower third: Counter-irritation to leg and foot: Rapid recovery.—James H., aged 37, came to hospital with considerable enlargement of the lower third of the left fibula. There was no pain in the joint. The use of iodine paint was maintained over the lower half of the leg, its whole circumference, and the posterior half of the foot. The symptoms disappeared with much more than usual rapidity.

Case LVI.

Nodes : A broad band of counter-irritation : Instantaneous subsidence of pain, and very rapid disappearance of swelling.—A node on tibia of two months' duration, and little benefited by iodide of potassium, the rest being greatly disturbed by pain. The application of a broad belt of iodine immediately relieved the pain, and the swelling very quickly subsided. A node of the clavicle had a history so similar as not to need repeating. *Remarks.*—The immediate disappearance of the pain in nodes is very constant. I presume this is, therefore, no coincidence.

Case LVII.

Ostitis of radius : Counter-irritation : Rapid recovery.—Thomas T., aged 22, married. For six months there had been an enlargement of the radius, which was gradually increasing, with pain and tenderness, but not in marked degree. There were no symptoms of disease in the elbow or wrist joints, but supination and pronation of the forearm were impaired

N

from the size of the bone. The integument of the whole of the forearm, and half the upper arm, were freely painted with iodine. In three weeks all symptoms had disappeared.

Case LVIII.

Ostitis of tarsal bones: Extensive and prolonged counter-irritation and rest: Complete recovery.—A boy, aged ten years, came to hospital with great swelling (but recent) of the tarsal bones of the left foot. Over the scaphoid and three cuneiform bones were four red patches, representing, very curiously, the shape and relations of the four bones. The skin of the foot, and a portion of the leg, was kept covered with iodine paint. Rest was secured by splintage and crutches. In six weeks, the inflammation had disappeared, but to ensure greater safety against relapse, the treatment was continued another three weeks.

Case LIX.

Ostitis of humerus: Counter-irritation: Rapid recovery.—I was consulted about a boy aged ten, who had enlargement and pain in the upper third of the humerus. There were no indications of disease of the shoulder point, beyond a little pain in movement. I recommended the application of iodine paint of a mild strength over the shoulder, the adjacent portion of the arm, and the adjacent portion of the thorax. The pain disappeared immediately, and in a few weeks the swelling also.

CASE LX.

Under Mr. Lloyd's Care—Necrosis of metatarsal bone of great toe : Singular benefit of counter-irritation. —I am indebted to Mr. Lloyd, of Ashted Row, for the report of the following case :—Alexander M., aged 37, when twenty years old, had slow ostitis and partial necrosis of bones of great toe, which required amputation, after thirteen years of every kind of treatment at every institution in the town. In November, 1869, he came under Mr. Lloyd's care with similar disease in the other foot. Mr. Lloyd applied iodine freely to the foot and leg up to the knee. The head of the metatarsal bone came away in six weeks, and the wound was firmly cicatrized on the 1st of March, 1870.

XII.

THE TREATMENT OF DISEASES OF THE JOINTS.

ACUTE synovitis, in my experience, is not a common disease, if we except the cases which are due to injury and pyæmia. When synovitis appears in conjunction with rheumatism, or gout, or syphilis, it is mostly

subacute or chronic. When strumous, it is essentially
chronic, and is generally, if not always, associated with
some degree, however slight it may be, of osteal inflammation.   I am compelled to differ from authors on this
point, but a careful examination of the cases of
so-called strumous synovitis, chronic thickening
(Brodie's degeneration) in the young adult, leads me to
the opinion that at least slight ostitis underlies and
probably precedes the synovial inflammation. A slight,
slow ostitis often requires to be carefully looked for,
while synovitis is obvious to the most uninitiated eye.
This slight ostitis appears to me to be insufficiently
recognized in the interpretation of many surgical
cases.   Often, and where there is no synovial, or
thecal, or bursal, or ligamentous, or fascial disease, a
little impaired mobility, or a little weakness, or a little
pain, or a little obscure swelling, simply or together,
are complained of for a few weeks or months, and then
completely disappear.   In these cases, if they are not
passed over as too trivial, evidence will commonly be
found of some bone inflammation, however slight.   If
now to one of these cases chronic synovial inflammation be added, the ostitis is lost sight of, or if, after a
time, bone symptoms unmistakeably manifest themselves, it is then said that the synovial disease has
extended to the bone.

   The vast majority of joint diseases, from the slightest

to the severest, have, I believe, an osteal origin. The severity of the joint disease also depends on the degree of ostitis. If the ostitis be slight or moderate, it and its attendant, articular disease, are mostly not difficult to cure. If it advances to caries, the articular symptoms are correspondingly more acute, and the treatment is more difficult. In this place I can only throw out these few suggestions on a subject upon which, at another time, I may have much more to say.

SYNOVITIS.

IN speaking of the more acute varieties of synovitis, excluding at present those cases where suppuration is present, the treatment to be described refers to the knee. In these cases, as an active, quickly acting, and superficial counter-irritant, nitrate of silver, in a strong solution, is very convenient and singularly successful. The front and sides of the knee, and a third of the anterior aspects of the thigh and leg, should be moistened twice or thrice with moderate freedom. The joint should be compressed with fine cotton wool and a bandage, and the limb fixed with a light gutta percha or other splint,* and at the same time a little

---

* I find a very convenient mode of applying a gutta percha splint for any purpose, to be this : The gutta percha having been cut something larger (because of contraction in hot water) than the required shape, is enveloped in two layers of calico sewn at the margins. It can then, by immersion in boiling water, be made as soft as dough, and can be readily applied without being pulled out of shape or sticking to the fingers. Complete softness when applied is the secret of success.

elevated. Rest in bed is of course necessary, but the diet need not be altered, except in moderating or stopping stimulants, and no medicine is required. With the treatment just laid down, the swelling mostly subsides in twenty-four hours.

In cases where *suppuration* is clearly present, free incisions, in a dependent position, should be made without delay, but in other respects the above treatment ought to be carried out. The counter-irritation should be a little more vigorous and a little more extensive, because of the greater severity of the disease. A poultice may also be substituted for the cotton wool, but it should not be applied immediately, as its moisture dilutes, washes off, and weakens the counter-irritant. In cases of suppuration, the iodine liniment may be used in place of nitrate of silver.

In the treatment of acute rheumatism, it appears to me that nitrate of silver would be better than blisters, it is more superficial and much quicker in its action.

In *chronic synovitis*, the treatment differs from that of acute inflammation, in being less vigorous. A mild iodine paint may be used, but it should be applied to a third or half the thigh and leg and to the whole circumference. Pressure is of the greatest use in this as in other chronic inflammations, as well as rest by means of light splintage.

ARTICULAR OSTITIS.

I WILL suppose that we have first to deal with a case which is not severe, and is in its earlier stages. The treatment is very similar to that of ostitis at the extremity of the bone where the joint is not involved, and which has been already described. Complete rest, with splintage, is here absolutely indispensable. The iodine liniment freely applied around, above, and below the affected joint, need only be repeated every two, or three, or four days. Pressure and elevation are of great service.

If the case be more advanced, or if caries threatens, or has already set in, but not to so persistent a degree as to require amputation, the treatment just referred to will often suffice to remove the disease, and in a much shorter space of time than when rest only is resorted to. But in these cases, and especially in thickly covered joints, the actual cautery possesses great advantages in its greater, deeper, and more persistent action, it has a drawback, however, in the subsequent suppuration which necessitates somewhat frequent dressing and consequent unrest. The actual cautery is now used by many surgeons, especially in hip disease, but in far too limited a manner and too near the joint. I shall speak of hip disease separately. In the shoulder, a stripe of eschar, half an inch wide,

should commence high in the axilla, and be carried
six or eight inches along the axillary and brachial
artery; two others should be made, one along the
anterior, the other along the posterior, aspect of the
joint, in such manner that the three stripes are equally
distant from each other. If the knee is affected, four
stripes of eschar should be made along the front, back,
and side of the joint. These should be ten or twelve
inches long, and also equally distant from each other.
The anterior stripe being carried over the quadriceps and
patella, and the two lateral ones over the condyles, a too
near approach to the synovial membrane is avoided. If
the surgeon prefers it, the posterior stripe may be made
broader and the others narrower. The spaces between
may be covered with iodine liniment. Gutta percha, or
other apparatus, to secure rest is quite indispensable,
with moderate pressure and elevation.

In the great majority of cases, the milder treatment
will probably be sufficient. In the exceptional cases, if
it be objected that the treatment is severe, it must be
remembered that the alternative is amputation   One
thing I cannot too emphatically say, namely, that this
treatment will rarely fail to cure every case in which excision of the knee is performed by some surgeons;
indeed, it renders excision, *with few exceptions*, an
unnecessary operation. The elbow is an exception chiefly
because here excision is performed to secure mobility,

not because the disease is incapable of cure by other means. In the hip, knee, shoulder, and wrist extensive (if need be) counter-irritation is surely preferable to excision. Cure with anchylosis in the shoulder matters little because of the mobility of the scapula. In the hip, excision, though it may be desirable in the last stages of disease, is not a very successful operation. In the wrist, the free use of iodine to the hand, and forearm, and wrist, rarely fail, and the moderately free use of the actual cautery will leave few opportunities for excision.

Whether anchylosis follow the treatment of articular disease or not, depends on the degree of inflammation of the bones and cartilages and synovial membrane. Rest should be resorted to in every case; if the inflammation be slight, mobility will be regained, or will be restored by the surgeon. In more advanced disease, anchylosis is the best attainable result.

In disease of the *hip-joint* I have found the most satisfactory results from the following treatment: A broad stripe of eschar is made over the femoral artery, from the groin to the lower third of the thigh, and another posteriorly, from near the crest of the ilium to the middle of the thigh. The limb is then (with considerable force, if necessary) restored to a good position. A well-padded long splint will best maintain a straight position. In my cases a pillow is

o

invariably placed transversely under the pelvis, so that
not only is the hip elevated, but the weight of the
limb constantly tends to overcome the flexion of the
thigh.

In *sacro-iliac* disease the same principles should
regulate the treatment, excision being imprac-
ticable.   In these cases there are the strongest
reasons for the effective use of iodine in early and
mild disease, and of the actual cautery in more marked
and advanced disease.   Rest to the joint and the
adjacent articulations must be protracted and complete.

What of drugs in inflammations of bones and joints?
Inflammation in bone does not differ from inflammation elsewhere—hence if any drug is useful in any
inflammation, let it be given here.   Because I believe
iron is advantageous in all inflammations, I give iron
in all osteal inflammations.   Mercury, especially the
bichloride, or the iodide of potassium, may be added
to the iron.

The inflammatory character of chronic rheumatic
arthritis was doubted by Dr. Todd, and, while not as-
senting to this view, there certainly is a peculiarity
in the pathological process which renders it little
amenable to the remedies for inflammation.   Even rest
here, especially if the disease has made any progress,
is of no service.   Iodine liniment, or flying blisters,
give more relief than any other remedies.

WOUNDS OF JOINTS.

SUPPOSING, as is usually the case, the knee is wounded, it should be carefully closed with silver sutures, a blister (12 by 3) should be placed along the femoral artery, the rest of the front of the thigh and leg covered with iodine liniment, the limb secured with light splintage, and the knee firmly compressed with cotton wool. Even if suppuration should occur, no other treatment promises so much benefit. A stronger measure, acting partly as a line of blister, is the ligature of the femoral artery. If suppuration were uninfluenced by counter-irritation, I should resort to it. In a suppurating joint from wound, I see no reason why a few long stripes of actual cautery should not be beneficial. As I have said elsewhere, why relinquish our more powerful remedies when the inflammation becomes severer? Incisions may also become necessary.

INFLAMED BURSÆ, SHEATHS OF TENDONS, AND FASCIÆ.

IN their earlier stages, inflamed bursæ may generally be cured by counter-irritation, if sufficiently extensive. In housemaid's knee, an oval patch of iodine (liniment), nine inches by six, or a blister, five by three, followed by iodine if not itself sufficient, will usually remove a

bursal fluid collection of recent formation. A circle
of blister or iodine answers as well. Careful pressure
should also be maintained during, and for some time
after, the treatment. All other early bursal inflammations are best treated in this manner. Bunion even,
if rest cannot be secured, is greatly relieved by
counter-irritation.

The longitudinal tender swellings, caused by
inflammation of the synovial sheaths of tendons, and
which are found mostly in the vicinity of the wrist
and ankle, are most successfully treated by a couple of
stripes of counter-irritation, by means of iodine, on
each side. A single broad stripe of nitrate of silver,
over the swelling, answers very well, especially if the
inflammation is recent, and if there is little prospect
of repeated irritation being necessary. Moderate
pressure and rest to the adjacent joint are also useful.

Inflammation of the fasciæ, if unchecked, gives rise
to contraction. It is frequently due to injury or
peculiarity of position in certain occupations. Iodine
paint, over a comparatively large surface of superjacent
skin, is a remedy which will rarely fail. In ulceration
or rupture, the iodine should be applied in the form of
a belt.

Case LXI.

*Articular ostitis of elbow : Iodine counter-irritation to
arm and forearm : Rapid recovery.*—Maria B., aged

28, came to hospital with enlargement of lower extremity of humerus, with impaired and painful movement of the elbow, of six months' duration. A light gutta percha splint was applied to the bend of the elbow, after two-thirds of the arm and two-thirds of the *forearm* had been freely moistened with iodine pigment. Directions were given that iodine should be applied every second day. In a week, the movement of the joint could be made without pain. In six weeks, the recovery was complete.

CASE XLII.

Articular ostitis of ankle : Counter-irritation to leg and foot : Good recovery.—James H., aged 37, came to hospital with enlargement of lower extremities of the left tibia and fibula and pain in the ankle joint. Free counter-irritation to the leg and foot, light splintage and crutches were ordered. In seven weeks, recovery was complete, with the exception of the enlargement, some of which remained and would probably be permanent.

CASE XLII.

Severe ostitis of knee : Counter-irritation to the thigh and leg : Speedy and complete recovery.—Mary C., aged 12, came with marked and advancing symptoms of articular ostitis of the knee. There was swelling of the femoral condyles and tibial tuberosities, wasting of the whole limb, starting pains, and contraction of the hamstring muscles. The disease was of seventeen months' duration, and the patient had formerly been in the hospital. Counter-irritation was effected by means of iodine to the thigh and leg, and maintained by removing the light splintage every third day. Recovery was complete in one month.

Case LXIV.

Ostitis of wrist : Counter-irritation to whole of forearm and hand : Subsidence of active symptoms in a week : Complete recovery in a month. — Mary B., aged 15, came to hospital with swelling, pain, tenderness, and impaired movement of left wrist. There were occasional, though not severe, starting pains. The forearm was somewhat wasted. The forearm, to the elbow, and the hand, were freely covered with iodine paint (a strong one, so that less frequent removals of splintage might suffice). In less than a week, pain and tenderness completely disappeared. In a month, recovery was complete. *Remarks.*—The wrist, however much diseased, is very amenable to counter-irritation. This is the more satisfactory because excision of the wrist, for several reasons, will never become a common operation.

Case LXV.

Hip disease, with large abscess: Counter-irritation: Rest: Recovery in four weeks.—Charles B., aged 4, came into hospital with all the symptoms of articular ostitis of the hip. A large abscess existed in the gluteal region. Iodine was applied for some distance around the joint, and rest secured by splintage. In a few days, the abscess opened, and closed in a few days, and in four weeks, all symptoms had disappeared. He was kept in bed a few weeks longer, for the sake of safety.

Case LXVI.

Advanced caries of the wrist, with great swelling and old sinuses on dorsal and palmar surfaces: The actual cautery in four long stripes: Recovery in one month.— Eliza E., aged 30, married, looked 50, came into hospital with caries and sinuses of left wrist, of old standing (sinuses 12 months, ostitis several years). Four stripes of eschar ($\frac{1}{2}$ inch wide) were made with the actual cautery, from the middle of the forearm to the clefts of the fingers, one at the dorsal surface, another at the palmar surface, another at the radial border, and another at the ulnar border. In one month, the symptoms had subsided and the sinuses had healed. *Remarks.*—This case was regarded as inevitably doomed to excision. With sufficient counter-irritation, I believe excision will very rarely be necessary. Rest *alone* had failed completely.

Case LXVII.

Ostitis of bones of both knees: Iodine to thighs and legs: Speedy recovery.—Alfred S , aged 10. After an injury of which only a vague account could be given, both knees became the seat of pain, osteal enlargement, and contraction of the flexor muscles. Both knees were straightened without much difficulty, and kept so by means of splintage. Iodine was freely applied every third or fourth day to the thigh and leg, the halves adjacent to the knee. In a fortnight he was sent out of the hospital quite well.

Case LXVIII.

Hip disease: The actual cautery in a long stripe at the front and back of the joint: Good recovery, followed by ostitis of wrist: Iodine to wrist and forearm: Recovery.—Sarah A. H., aged 13, came into hospital with severe symptoms of hip disease, of 12 months' duration. Her general health was quite broken down. The actual cautery was applied so as to produce two stripes of eschar ($\frac{3}{4}$ inch wide), one at the front and upper two-thirds of the thigh, over the femoral artery, another at the back, from a few inches above the joint to the middle of the back of the thigh, passing between the trochanter and ischial tuberosity. Rest was secured by a splint. The joint symptoms quickly subsided, but the general health only slowly recovered. Three months after admission, the left wrist suddenly became the subject of ostitis, for which the hand, and part of forearm, were covered with iodine. She left the hospital cured, in four months. *Remarks.*—Recovery here was longer than usually required with this treatment. The profound struma explains this.

Case LXIX.

Acute synovitis of knee from injury : Subsidence in twenty-four hours.—A medical gentleman brought his right knee into forcible conduct with a piece of iron. In a few hours the joint was much swelled, with pain, tenderness, and fluctuation, but without marked constitutional disturbance. The joint, and portion of thigh and leg, were freely moistened with a strong solution of nitrate of silver; cotton wool and a bandage were then applied. In twenty-four hours, the swelling disappeared. *Remarks.*—The treatment of recent and acute, or subacute synovitis, with arg. nit. applied not only on the knee, but the thigh and leg, for a third or half their length and their whole circumference, is rapidly successful. In more chronic cases, with thicker coverings, iodine is perhaps preferable.

P

Case LXX.

Old ostitis of wrist: Actual cautery: Complete recovery.—Mary B., aged 25, of intensely earthy complexion, came with symptoms of old and severe ostitis of right wrist. The forearm was wasted, starting pains were frequent and severe. The general health was failing. The disease had made constant progress for eighteen months. The actual cautery was applied in stripes over the wrist. The pain immediately ceased, and the symptoms entirely disappeared (although more slowly than usual) in eight to ten weeks.

Case LXXI.

Bursitis (patellar): Counter-irritation: Complete recovery in ten days.—Mary Ann B., aged 30, came to hospital with house-maid's knee, which she had noticed for many months. The swelling was prominent but not very tense, and the cyst walls were not very thick. An oval blister (4 by 3) was put over the enlargement. In a week, recovery was complete. Iodine paint was tried before the blister, but was not sufficiently powerful. *Remarks.*—It may be said with certainty that, short of operative measures, counter-irritation only would have cured the above case. Can as much be said of any other remedy in any other inflammation?

Case LXXII.

Bursitis (patellar) : Counter-irritation : Recovery in a week.—Thomas P., aged 33, labourer, came into hospital with enlargement and fluctuation of the patellar bursa, of six weeks' duration. No history of injury. Was increasing in size up to moment of admission. The pure iodine liniment was painted over and around the swelling, along the thigh above and leg below. The application was repeated daily for a few days, and then every other day, for a few days longer. In forty-eight hours, the swelling had nearly disappeared, in a week every vestige of disease had gone. *Remarks.*—If the disease had been a little more advanced here, stronger counter-irritation would probably have been needed.

Case LXXIII.

Inflammation of sheath of peronei tendons : Broad stripe of counter-irritation : Rapid recovery.—A young man, who had sprained the left ankle many weeks before, applied with a longitudinal swelling over the peronei, at the back of the fibula, and obliquely forwards over the calcis. Iodine liniment was applied in two vigorous stripes, one on each side of the tender and painful swelling, with a milder one between the two. In a few days, all the symptoms had disappeared.

Remarks.—Of course, the treatment by counter-irritation of thecal and bursal inflammations is not new. My mode of treatment here embraces and modifies what is old. What should we do without counter-irritation in such cases? Three days' counter-irritation does more good than three months' rest.

Case LXXIV.

Inflammation of sheath of extensors of metacarpal bone, and first phalanx of thumb: Counter-irritation: Immediate recovery.—A middle-aged woman came to hospital with a tender, longitudinal swelling at the outer border of the forearm (near the wrist), and of the wrist, and thumb, of two months' duration. Two vigorous stripes of iodine, and a mild intervening one, removed all the symptoms in a week.

Case LXXV.

Inflammation of bicipital fascia, from occupation: Counter-irritation and rest: Rapid recovery.—Matthew J. R., 27, miner. Right elbow constantly flexed more or less in work. For eight weeks before admission, elbow firmly flexed, with tenderness and rigidity of fascia in bend of elbow. A broad circle of iodine irritation was used, and the arm supported. In eight days, recovery was almost complete.

XIII.

THE TREATMENT OF MUSCULAR INFLAMMATIONS.

DISEASES and injuries of the muscles are conditions which for the most part require operative or mechanical treatment. Inflammation of muscles, as indicated by circumscribed swelling, induration, tenderness, and impaired use is commonly, and I believe justly, considered to be syphilitic. Any muscle may be affected, but the sterno-mastoid more frequently than others, and especially in infants.

In the treatment, a mild counter-irritation is by far the most valuable local remedy. Laborious attempts at rest are scarcely necessary, and in some cases, as when the sterno-mastoid is affected in children, impossible. In children, the counter-irritation should be guardedly mild,—the tincture of iodine only being used. Internally in adults, the iodide of potassium will usually suffice; in children, however, the grey powder is the best remedy.

In severer muscular inflammations in the adult, especially where there is ulceration, or abscess, or sloughing, or gangrene, the counter-irritation should be in larger zones, and perhaps more vigorous in character. Save in the very young, cantharides may be substituted for iodine, but of course, the zone must be of diminished width. In such cases, too, as much rest must be secured as is practicable.

Sprains may perhaps be considered here as appropriately as any where. Those of the *ankle* are the most frequent and the most formidable. The ankle should be immediately compressed by strapping with adhesive plaster. Just above the strapping, a ring of counter-irritation (with iodine liniment in the first instance) will not only retard and diminish subsequent inflammation, but will also relieve the severe pain. Another successful method, in the ankle and other joints, is first to compress with a domette bandage, and then outside this to apply very hot fomentations. Whatever joint be affected, it is most important that it be elevated for a time and completely rested.

Case LXXVI.

Bruise and Inflammation of muscles of leg: Counter-irritation: Rapid recovery. — Richard B., aged 42, after a severe injury to the muscles of the back of the leg, there remained, after several weeks, irregular indurations in the substance of the muscles of the calf, with much pain on extending the ankle. A broad stripe of iodine liniment was applied to the *front* of the leg. In four days, there was great improvement, and in another few days, recovery was complete.

Case LXXVII.

Inflammation of sterno-mastoid in a young child : A horseshoe of iodine irritation : Immediate recovery.—**A** gentleman consulted me about his infant, a few weeks' old, whose left sterno-mastoid was the seat of contraction, and of a hard and painful swelling. Hereditary syphilis was probably present. A horseshoe patch of skin was moistened with a weak iodine paint. In forty-eight hours, the disease had nearly disappeared. A small dose of chalk and mercury was also given each day, and continued for some time. *Remarks.*—This is the type of a not rare class of cases.

XIV.

Treatment of Inflammatory Diseases of the Vascular and Absorbent System.

Phlebitis, as the German pathologists have shown, is an inflammation altogether secondary to thrombosis or embolism; it may be treated locally by stripes of counter-irritation.

Inflammation of the lymphatics, as indicated by the red lines in the superjacent skin, is a condition which does not require treatment so much as the prior and severer disease upon which it depends. If it should

seem to possess any independent importance, a stripe
of skin may be painted with a strong solution of
nitrate of silver.   If the glands to which they lead
are also inflamed, the same solution should be carried
over them, and for some distance around them.   The
treatment for suppurative inflammation, in or around
the glands, is similar to that described in the next
paragraph.

Inflammation of the lymphatic glands is one of the
commonest of surgical diseases, and may present any
stage of severity, from the most acute to the most
chronic inflammatory action.   The great majority of
cases, however, are those in which the inflammation
is very chronic, and present themselves as cases of
" enlarged glands."   It is of the treatment of these
cases that I shall now speak, as the treatment of acute
suppurative inflammation of the glands is the same as
that of abscess.   Indeed, the difference between the
treatment of any inflammatory diseases is one of
detail—detail, however, of no slight importance.   A
broad zone, or semicircle, or adjacent patch of iodine
paint is the one essential feature of the treatment
The paint should be moderately strong, or the liniment
may be used where the skin is thick.   In the very
young, the paint should be correspondingly weaker.
In these cases it is desirable that no paint should be
applied over the glands.   Enlarged glands, as every

surgeon well knows, are often extremely obstinate, and refuse to yield to any treatment hitherto known. I will not say that the treatment described here will never fail, but it has not failed in one of my cases. It does not cure them in a few days, as it often does an acute inflammation, but it removes them in a few weeks, it may be a few months, when the accepted modes of treatment have been tried for years and failed. I can scarcely doubt that it will become the common treatment for enlarged glands until a better treatment be devised, whether that be soon or late.   Pressure with a shot bag for limited periods during the day, the patient lying on a couch or bed with the glands upper-most, is also of service.   Internally, I give iron, or iron with iodide of potassium.   I have never been able to trace any benefit to cod liver oil.   Local treatment, in the cases described, is more important than the general treatment.   It is the reverse with enlarged glands from acquired syphilis.

Case LXXVIII.

Enlarged " pectoral" glands : Counter-irritation over the brachial artery : Immediate benefit.—Miss O. L. was brought to me by her mother, at the advice of the family medical attendant, for enlargement and tender-ness of the glands, on the border of the large pectoral muscle, between the mammary gland and the axilla.

I directed a stripe of counter-irritation to be maintained over the inner side of the arm. The two ladies called upon me eighteen months after, stating that the enlargement disappeared immediately on the application of the paint (a rather strong one); that thrice subsequently a little enlargement and tenderness had reappeared, but was immediately removed each time on the re-application of the paint.

Case LXXIX.

(*Under Mr. Hickenbotham's care*). *Great enlargement of the cervical glands: Counter-irritation at the back of the neck: Recovery in three weeks, after three years' failure with every other known treatment.*—Mr. Hickenbotham has informed me of the case of a young man, with great enlargement of the cervical glands of three years' duration, who had been to many surgeons, and tried every known treatment without benefit. Mr. Hickenbotham kept up a patch of iodine irritation at the back of the neck, and in three weeks the whole enlargement disappeared. Mr. Hickenbotham had previously tried many other remedies.

Case LXXX.

Enlargement of cervical glands, of many years' duration: Circumjacent counter-irritation: Subsidence in one month.—Alice L., aged 15, had great enlargement of cervical glands many years. A large patch of iodine paint was made at the back of the neck, and continued below the glands in a curved manner. A shot bag was directed to be worn over the glands occasionally. In one month, the enlargement disappeared. Every kind of treatment had been tried before without success.

Case LXXXI.

Strumous enlargement of glands in groin: Counter-irritation: Rapid improvement.—George W., aged 13, enlargement of cluster of inguinal glands, of long duration, associated with strumous ulcers. Iodine paint was applied in the form of a horseshoe, and the glandular enlargement disappeared very rapidly. The strumous ulcers were much more obstinate—they had very little inflammatory action about them.

Case LXXXII.

Great enlargment of cervical glands, of three years' duration, and resisting every treatment: Adjacent counter-irritation: Rapid subsidence.—Emma H., aged 15, had enlargement of cervical glands for three years, and tried all kinds of treatment without success. A large disc of iodine was applied to the back of the neck, and the enlargement very rapidly subsided.

Case LXXXIII.

Enlarged cervical glands: Long duration: Counter-

irritation : Rapid improvement.—James D., aged 33, had strumous enlargement of the glands many years. A crescent of iodine irritation removed the swellings in three weeks.

Case LXXXIV.

Enlargement of cervical glands of long standing : Counter-irritation : Immediate improvement.—Ellen R., aged 9, came to hospital with old-standing enlargement of cervical glands, on which much and varied treatment had been expended in vain. A horseshoe of iodine paint was applied around the swelling, with immediate improvement of a very marked character. The subsidence was very rapid until very little enlargement remained, when the progress was not so rapid. *Remarks.*—Probably the treatment was carelessly carried out towards the end.

XV.

Treatment of Inflammatory Diseases and of Injuries of the Nervous System.

In limited, and not very severe inflammations of the *scalp*, abscess, or ulceration, or carbuncle ; in limited inflammations of the *periosteum*, acute or chronic, and in limited *ostitis*, or *caries*, or *necrosis*, the hair should be kept closely clipped, and zones of counter-irritation

effected by means of a strong solution of nitrate of
silver. When counter-irritation is necessary over the
scalp, it should be very superficial, particularly in
children; hence in such cases the nitrate of silver is
most applicable and most successful.

Where the inflammatory condition is more extensive, whether in the scalp as abscess, or carbuncle, or
erysipelas, or cellulitis, or in the periosteum, or bone,
as extensive periostitis, or necrosis, the counter-
irritation should be along the anterior borders of the
sterno-mastoids (one or both), over the carotid vessels.
The stripe of artificial inflammation may be produced
by iodine liniment or cantharides in the more acute,
and by an iodine paint in the more chronic
inflammations. Any uninvolved portion of the
scalp may also be painted with nitrate of silver
in the acuter cases, where efforts to cut short the
inflammation are so much more successful. A large
patch of iodine at the back of the neck answers very
well in every inflammation of the head or face.

The treatment just described is also precisely the
best treatment in the *intra-cranial inflammations* which
come before the surgeon—those, namely, which follow
injuries of the head. After concussion or contusion
of the brain, or fracture of the skull, febrile symptoms,
especially an elevation of temperature as indicated by
the thermometer, should call forth active treatment.

The counter-irritation not only does not interfere with elevation of the head and shoulders, mercurial inunction, or active purgation, but is actually assisted by those measures. A common remedy—cold to the head —I have never known to be of any service : by diminishing the external, it can only increase the intra-cranial circulation. In its stead, the nitrate of silver may be advantageously applied to the shaven scalp.

In *syphilitic inflammation of the cerebral membranes, or of the brain surface*, the same treatment is also the best. Here, and in all intra-cranial inflammations, the disc at the back of the neck, or the stripe of counter-irritation, from the sternum to the mastoid process, should be especially widened and intensified over the mastoid process, where, for obvious anatomical reasons, it must, from every point of view, be most efficient. At the mastoid process the two circulations, extra- and ultra-cranial, communicate most freely, while underneath and below it are all the great vessels and nerves which are connected with the head.

In the comparatively rare inflammatory affections of the *nerve trunks*, whether due to injury or disease, there is no local remedy of any service which does not act as a counter-irritant. How the counter-irritation shall be produced, and how much of it, must be determined by the depth and locality of the nerve. In such cases, rest should form an important item in the treatment.

Inflammations arising from injuries to the spinal column.—When symptoms of inflammation of the cord or its membranes occur after *concussion*, or *contusion*, or *compression*, or twists, or sprains, they are extremely persistent, and very often progressive, in spite of all treatment. Counter-irritation is often resorted to in such cases, but in an inefficient and perfunctory manner. If such cases are not quickly benefited by two long wide efficient stripes of iodine (liniment) irritation, the actual cautery should be used on both sides of the affected structures, and, seeing the importance of the issues, in a manner widely different from its ordinary niggardly use. I frequently produce longer and broader stripes of eschar by the actual cautery, in cases where, if they fail, there is no other remedy left, than any other surgeon, and in no single case have I seen shock, or other ill, result. Indeed, in no single case have I seen anything but unquestioned good, if not, and this not very rarely, ultimate cure. Rest, in all spine diseases, is indispensable. The bichloride of mercury is often given internally, with what benefit I am not able to say.

Caries of vertebræ, whether with or without angular curvature, at whatever age, in whatever locality, at any stage, should be treated by the counter-irritation of the actual cautery. The disease is deep, protracted, and formidable. The counter-irritation should be deep and protracted—formidable it happily is not with the beneficent aid of chloroform. Two stripes of eschar, an inch wide and twelve inches long, one on each side of the disease, would not be too extensive (considering the alternative) in an adult. In the young, the eschar should be proportionately less. In disease of the upper cervical vertebræ, the eschar need not be so long; it may be carried upwards nearly to the upper occipital ridge, and downwards to between the shoulders. Should counter-irritation be rejected after suppuration has occurred? I have already expressed my opinion on this question. I have lately had an opportunity of using the cautery in lumbar abscess, with a splendid result. Unless death were close at hand, I should certainly resort to its free use, as I should have it used on myself if I had a spinal abscess, and its almost inevittable prospect. The abscesses themselves are, I need scarcely say, not treated as primary local abscesses. On the need of absolute rest it is impossible to speak too strongly. The diet should be nutritious, with little or no stimulant, unless there be profuse suppurative discharge. Opiates may be needed, and iron, alone or with the iodide of potassium, or the bichloride of mercury may be given. In caries of the spine, however, as in most inflammations, the surgeon should have two great questions before him : how best to establish counter-irritation, and how best to secure rest. If these two arrows fail, he has none left in his quiver.

CASE LXXXV.

Angular curvative (caries) in the dorsal region, with cord symptoms: two long stripes of actual cautery : rest : cure in two months.—Elizabeth G., aged 18, came into the hospital with angular curve in the vicinity of the fourth dorsal vertebra. There were indications in the motion and sensation of the lower extremities, and in the bladder, of implication of the cord in the inflammatory disease. I made two stripes of eschar with the actual cautery, twelve inches in length and one in width, on each side of the curve. Rest in bed (trebly inclined plane) was of course maintained. In two months she left the hospital quite well.

CASE LXXXVI.

Angular curvature (caries) of the spine in the lumbar region, of twelve months' duration : the actual cautery in two long stripes : rapid recovery.—Joseph G., aged 12, came into hospital with an angular projection in the upper lumbar region, of twelve months duration. Two long stripes of eschar were made on either side the curve. In a month, all the symptoms had disappeared, and that, too, in spite of the utmost restlessness in bed. He was kept in bed some time longer, for safety's sake.

R

Case LXXXVII.

Angular curvature with lumbar abscess: Two long stripes of actual cautery: Complete and rapid recovery. —A boy was admitted into hospital with marked angular curvature in the lower dorsal region, and a large lumbar abscess. Two long eschars were made with the actual cautery. In a few days, the abscess opened; three weeks later, it was quite closed. In six weeks, recovery was complete.

XVI.

The Treatment of Inflammatory Diseases of the Respiratory Organs.

Diseases of the nose. — In the treatment of inflammatory affections of the nose and mouth, it is a disadvantage that counter-irritation cannot be established in the face, at least without so much unsightliness that only urgent reasons can justify it.

Such urgent reasons certainly exist when *specific inflammation* threatens to destroy, or is actually destroying, the nose. Here, a horseshoe of counter-irritation should be established over the cheeks and upper lip, by means of strong acetum lyttæ, and maintained as long as necessary by preparations of either cantharides or iodine. This treatment must be associated with constitutional remedies, such as opium or mercurials, or twenty grain doses of iodide of potassium.

The great majority of nasal inflammations are of a more chronic character, and the counter-irritation need not be in a conspicuous locality. In *periostitis of the nasal bones,* in affections of the nasal cavities, as *chronic inflammatory thickening of the mucous lining, ozœna*— strumous or syphilitic, *ulceration,* or *abscess,* and in repeated *epistaxis* counter-irritation may be effected at the back of the neck. This might be a transverse oval patch or half collar of iodine irritation, or, what is simple, easy to manage, and not admitting of neglect, a thread seton. Chronic thickening and ozœna are so obstinate that they may be said to be without treatment, unless counter-irritation be a remedy. The direct application of lotions I have never known to be of benefit. Iron internally, or iron with iodide of potassium, or minute doses of mercury, are of possible, nay, of probable, benefit. However small it is, it is worth while to obtain it.

DISEASES OF THE LARYNX.

IN *laryngitis,* acute or chronic, in *œdema,* in *ulceration* of the larynx, in *necrosis of the laryngeal cartilages,* and in inflammatory conditions of the epiglottis, counter-irritation may be most usefully established in two stripes over the carotid vessels, and may be produced by cantharides or by iodine. The width, the length, and the intensity of the stripe must be determined by the severity of the laryngeal disease. Many of the laryngeal conditions, especially inflammation, ulceration, and necrosis, are either specific, and require the addition of specific remedies to those just described, or they are associated with tubercular disease, and will require the general treatment of struma. Brisk counter-irritation will often avert the necessity of opening the air tube, but if the necessity should arise, it must be promptly met. Neither in practice nor in these pages do I discourage operative interference, whatever the nature of the case may be, where other treatment fails, and where an operation offers even slender chances of success.

Bronchocele, although not inflammatory, is treated with more advantage by counter-irritation at the back of the neck than by any other method. With this should be combined elastic pressure; a silk velvet band, with elastic webbing let in at the back, does very well.

Case LXXXVIII.

Bronchocele rapidly diminished in size by counter-irritation at the back of the neck.—Eliza W., aged 16, came to hospital with a moderately-sized bronchocele. A wide oval patch of iodine at the back of the neck reduced the bronchocele two-thirds in one month.

CASE LXXXIX.

(Kindly reported by Dr. Quirk).—Bronchocele cured by counter-irritation at the back of the neck.—E. H., aged 14 years, a healthy looking girl, consulted me in October, 1869, for bronchocele. The enlargement of the gland had been present for six months, and seemed to be on the increase. Ordered the enlarged gland to be painted every morning with tincture of iodine. This treatment was continued for some weeks with no apparent benefit. I then directed the whole posterior surface of the neck to be painted with the tincture. In a few days, there was a marked diminution in the size of the tumour. The treatment was continued, with occasional intermissions, for five weeks, at the end of which time the swelling had completely disappeared.

XVII.

THE TREATMENT OF INFLAMMATORY DISEASES AND OF INJURIES OF THE DIGESTIVE ORGANS.

IN speaking of the nose, I remarked that in its diseases the face is not a convenient locality for counter-irritation ; the same remark applies to the *lips*, the *mouth*,

and the *tongue.* Many, if not most, of the diseases of
the organs referred to are specific, and are to be cured
mainly by specific treatment. These, however, may
often be somewhat influenced, and simple inflammations may be considerably influenced by counter-irritation. It is still a question with me, in many cases,
which is the best locality, whether, for instance, the
carotids and mastoid process, or the sub-maxillary
region, or the back of the neck.

The mouth.—Cancrum oris, in its severer forms,
is now very rarely met with. If a case were to come
before me, I should paint a ring of integument *on the
cheek* and around the disease, with acetum lyttæ,
and a broader ring externally with iodine liniment.
The treatment, indeed, would be that of destructive
inflammation, to which the reader is referred.

The tongue.—In *acute glossitis* and in *abscess of the
tongue,* a blister should be applied, or the acetum lyttæ
be painted over the sub-maxillary region, and carried
backwards to the carotid, over which it may be
extended. In the more chronic inflammations, a
milder irritation may be applied, either in the locality
just described, or at the back of the neck, by means
of iodine. *Ulceration* of the tongue, when it is not
epithelial, is syphilitic, and is occasionally one of the
most obstinate of surgical diseases. A very mild
mercurial course and sub-maxillary iodine irritation is

what I have found to be the most successful treatment.
A surgeon was under my care, with syphilitic ulcers
of the tongue ; a physician told him they were due to
mercury ; under the influence of the sixth of a grain
doses of calomel he was speedily well. His wife was
under the care of the physician (not for the ulcers,
however), she had ulcers of the tongue, but took no
mercury. The ulcers simply grew slowly worse until
she was treated as her husband had been, when she
also rapidly recovered. Nodes, or " gummy " inflammations on the tongue are also best treated in the
manner just described. *Prolapse*, although probably
caused by inflammation, unless slight, will usually
require operative treatment.

The tonsil is rarely subject to any disease which is
not inflammatory. In *acute inflammation* and in *abscess*
of the tonsil, a blister or the acetum lyttæ should be
applied over the carotid vessels, and at the upper end
the stripe should be widened in a forward direction
underneath the ear and over the angle of the jaw.
In sub-acute inflammation, the acetum lyttæ is a
convenient remedy. I have repeatedly seen the sub-
acute varieties completely relieved, in twenty to forty
hours, by a small blister at the angle of each jaw.
In the chronic inflammation, and in the so-called
hypertrophy (which is inflammatory in its origin),
if it be not of long duration, the application of an

iodine pigment in the locality referred to is often very successful. In the ulcerations, simple or syphilitic, as also in diphtheritic disease, a similar treatment will usefully supplement constitutional remedies. It is most useful where the inflammation is most acute. The application of iodine or nitrate of silver to enlarged tonsils is often beneficial where the enlargement is not very great, and I have no doubt that such application acts as counter-irritation, as incisions do in erysipelas. The more effective this counter-irritation, the better the result. Thus a small eschar made on the surface of a large tonsil by *potassa cum calce*, is quickly followed by subsidence of the enlargement.

The Intestines and Abdomen.

In abscesses of the abdominal wall, whatever their origin, whether in the abdominal wall or in the viscera, or in adjacent structures, the safest principle which can guide our treatment is to diminish inflammatory action to the utmost. This is best effected by establishing, around the seat of disease, a broad zone of counter-irritation, by means of iodine, or acetum lyttæ, or nitrate of silver. If the abscess, threatened or actual, be near the groins, as is often seen in pelvic cellulitis, acetum lyttæ, or a stripe of blister, may be applied over the femoral artery, in addition to the zone or horseshoe on the abdominal wall.

Inflammatory conditions associated with hernia : in-flammation of the hernia, &c.—Every surgeon of experience is familiar with a condition in irreducible, and especially in umbilical hernia, in which, with the general symptoms of strangulated hernia, such as vomiting and constipation, the local symptoms are merely those of inflammation. In these cases, a circumjacent zone, or horseshoe, of counter-irritation produced by acetum lyttæ, is often marvellous in its result. Local rest by means of opium, and general rest in bed, are also of great importance.

In the *peritonitis* following or preceding operations for strangulated hernia, an effective horseshoe of acetum lyttæ around the seat of operation will be of, it is not too exaggerated language to say, immense service. Where, from lapse of time, and the condition of bowel discovered during operation, inflammation threatens a fatal result, it may, in my experience, be often averted by prompt, sharp, and properly situated counter-irritation. I need scarcely presume that the reader agrees with the modern abstinence from purgation, in the use of opium, and in scanty diet, all which details are simply carrying out the fundamental principle of *rest.*

The *peritonitis following injuries*, and operations on the abdominal wall, may probably be averted in many cases of wounds, or operations opening the abdominal

s

cavity, or if it has commenced it may probably be diminished, by the use of counter-irritation around the seat of injury. Extreme care in closing the wounds with silver sutures, together with opium and starvation, or support by rectum, are the other items of treatment. The rest prevents harm, the counter-irritation does actual good.

The *rectum* and *anus*.—I have just been describing conditions in which counter-irritation is of great, indeed often of magical, utility. I turn now to inflammatory conditions in which it is probably of little use, at least it promises so little I have not tried it. The reasons are obvious. Ulcer of the rectum, fissure of the anus, fistula in ano, are kept up by mechanical conditions, and are rarely important as inflammations. The removal of the mechanical conditions, the perpetual unrest especially, is so simple and so uniformly successful by the knife, that the surgery of the rectum, as regards all inflammatory diseases, is a series of triumphs to the surgeons. It is worth while remarking that in the pain of all diseases of the rectum, counter-irritation, in the form of very hot hip-baths, for short periods, gives the most relief.

Case XC.

*Syphilitic ulceration of the tongue, cheeks, and fauces :
Immediate relief on the application of counter-irritation.—*

A surgeon, who had been inoculated with syphilis through a scratched finger at a midwifery case, was suffering great annoyance from ulcers on the tongue, cheeks, and fauces. A single application of iodine paint externally, removed the discomfort immediately. Several of the ulcers healed in a few days, and all nearly so within a week. He was taking a little mercury at the time. *Remarks.*—The instantaneous relief forbids the idea of a mere coincidence.

CASE XCI.

Acute tonsilitis : Iodine liniment to neck and angles of jaws, with complete relief in forty-eight hours.—Sarah C., 21, came with all symptoms of acute inflammation of tonsils. Deglutition was very difficult and painful. The iodine liniment was freely applied to the neck over the carotids, and especially at the angles of the jaws. In a few hours, the relief was very marked; in forty-eight hours, it was complete. *Remarks.*—This is but one of many cases in which I have seen sudden and marked benefit on the application of counterirritation.

CASE XCII.

Inflammation of the sac of a large irreducible umbilical hernia, with symptoms of strangulation : Instantaneous relief from a circle of counter-irritation.—I was called to see, in consultation with her medical attendant, Mrs. N., aged 60. An irreducible umbilical hernia had been present several years. For last three days there were vomiting, constipation, and exhaustion. The hernial tumour was not tense or enlarged. In the wall of the abdomen, close to the tumour, and encircling its lower two-thirds, was a crescentic mass of inflammatory induration, which was tender and painful, and of the size of a man's open hand. A circle of acetum lyttæ was applied around the tumour, and opium given internally. All the symptoms quickly subsided; and in twenty hours, nine-tenths of the induration had disappeared. *Remarks.*—It may be said the opium stopped the vomiting, but opium never before in a few hours removed a large, hard inflammatory mass.

XVIII.

The Treatment of Diseases of the Urinary Organs.

The *Bladder.*—Inflammation of the bladder is perhaps always a symptom of some other disease, or an extension of some other inflammation. The success of the surgeon in treating cystitis will entirely depend upon the success which attends his efforts to discover the prior disease, and on the possibility of removing that disease when discovered. When cystitis depends on a cause that can be removed, as stone in the bladder, or stricture, the treatment is obvious. When it depends on causes that are persistent, as enlarged prostate, or paralysis, or atony, or tumours which prevent the egress of urine, the proper treatment is, as a rule, to secure the complete periodical emptying of the bladder; but added to these surgical measures, it is often advantageous to maintain a mild disc of counter-irritation by iodine, or the occasional application of nitrate of silver to a more limited extent.

In the cystitis which depends on the extension into the bladder of gonorrhœal, or other urethral inflammation, counter-irritation is essentially the best treatment; but before this is used with vigour, it must be made perfectly clear that there is no stricture from a previous gonorrhœa or gleet, if there be, the use of large bougies is of even more importance than the counter-irritation. Where a first attack of gonorrhœa implicates the bladder, the treatment would simply be a slight extension of the counter-irritation adopted for the cure of the gonorrhœa. The general treatment will be spoken of in the next paragraph. Diluents, alkalies, and hyoscyamus, or conium, are supposed to be of use.

In the cystitis of children, simulating stone in the

bladder, there is perhaps always a mechanical obstruction to the urinary flow, a phymosis, or a small meatus, or a congenital narrowing of some portion of the urethra. This should be relieved, if possible, for the persistence of the cause renders all other treatment nugatory.

The Kidneys.—In the inflammations of these organs, whether acute or chronic, whether due to injury, or the presence of calculi, or to impaired urinary egress, whether in the kidney, or the pelvis of the kidney, or the perinephritic tissue, counter-irritation promises more than any other treatment. Its extent and intensity must be determined by the severity of the symptoms. Iodine or nitrate of silver are the best applications. Dry cupping is less useful. When the inflammation is caused by obstruction to the urinary flow, caused by stricture, the stricture must be relieved; but any treatment of stricture, after renal disease has set in, is of a very unhopeful character. The skin and the bowels should also be acted upon, the first by means of warmth, and the second by aperients. Diluents are probably of service in both renal and vesical inflammations. In the use of vegetable infusions and balsams, each surgeon must use his own judgment.

The *Prostate.*—In *acute prostatitis*, the perineum, and the adjacent surface of the thigh, should be painted with a strong solution of nitrate of silver, and at the same time a stripe of counter-irritation may, with additional utility, be established over the femoral arteries, by iodine liniment, or even acetum lyttæ. Very hot hip baths for short periods, eight or ten minutes, as directed by Sir H. Thompson, add to the relief. Abscess of the prostate must be treated as acute inflammation. If spontaneous opening be deferred the surgeon must open the abscess in the usual manner.

In *chronic* prostatitis, the counter-irritation should be a little less extensive, whether nitrate of silver or acetum lyttæ or iodine be used, and may be confined to the perineum. Iodine, however, may be advantageously applied to the adjacent portions of the thigh, in addition to the perineum and the thighs. Much must be determined by the degree of chronicity.

The *Urethra.*—Cases of gonorrhœa and acute urethritis should (as I have found from considerable experience) be divided into two classes for purposes of treatment, namely, those in which the disease is present for the first time, and those in which it has occurred once or oftener before. The ground for the distinction is this : when the disease has been present before, we can never be sure that *some degree of stricture* is not present ; in other words, that a mechanical, as well as inflammatory condition, is not

present, and which will require other treatment than
that directed to the inflammation. Even first attacks,
when much prolonged, may require to be placed in
the second category.

Supposing the case before us to be a first attack,
and there is simply inflammation, no matter how acute
it may be, it may almost invariably be cured in *two or
three days* by counter-irritation. Acetum lyttæ should
be applied over both femorals, and with this should
be combined a disc, or broad zone, of iodine liniment
applied daily. A larger surface of iodine liniment
will suffice. The diet should be free from stimulants,
and an alkaline drink may be taken before meals. If,
after scalding and other active symptoms have passed
away, there remains a little thin discharge, the iodine
zone should be maintained for a short period. The
advantages of counter-irritation have already been
demonstrated more than once. The failures in
stricture cases have probably prevented its wider use.

If the attack is not the first, but a third or sixth,
the counter-irritation should be less active, and when
active inflammation has subsided, a large bougie should
be passed every two or three days. In obstinate cases
an astringent injection, chloride of zinc (1 gr. to 1 oz.)
being as good as any, may be added to the previous
treatment. In such cases, where there is some
organic change in the mucous and submucous tissues,

internal remedies are useless. Purgation (counter-irritation to intestinal canal,) is often of use in inflammation of the urinary organs.

In *gleet*, and *chronic urethritis*, and *incipient stricture*, moderate counter-irritation, with iodine, large metal bougies, and perhaps astringent injections are the main remedies. In balanitis, moderate counter-irritation to the body of the penis (this may also be resorted to in urethral inflammation in addition ·to what has been described), and a little in each groin is the best remedy.

Stricture of the urethra, as a rule, unquestionably requires surgical treatment, and the reader is referred to the numerous and able works on a subject to which surgeons are justly giving so much attention. It should not be overlooked that the use of instruments is not, in every case, and at all times, to be resorted to. Rest, warmth, and counter-irritation are agencies which the surgeon will sometimes find useful, and very frequently useful, in combination with the careful use of instruments. *Urinary fistula and urinary abscess*, although inflammatory conditions, are, nevertheless, the result of surgical conditions requiring surgical treatment.

The Testis.—The treatment of *acute orchitis*, according to the principle enunciated in these pages, has been already brought under the notice of the profession. Orchitis has often been cured in twenty-four hours

T

by the following method : The scrotum is painted with
a strong solution of nitrate of silver, and a broad
stripe of skin over the femoral artery, of the same
side, is freely covered with iodine liniment; the scrotum
should be covered with cotton wool, and moderately
compressed　.No drugs need be given.

In other varieties of orchitis, such as the syphilitic,
the tubercular and the chronic, the action is so extremely chronic (more like the development of a
tumour than inflammation) that counter-irritation is of
little use.

The scrotum.—In the majority of inflammatory
enlargements of the scrotum there is usually some
causal stricture and urinary infiltration, to which
treatment must be directed. In exceptional cases,
erysipelas, œdema, cellulitis, and gangrene, attack
the scrotum as *apparently* primary diseases. Active
counter-irritation, according to the intensity of the
disease, should be the principal treatment—acupunc-
ture or incisions being added if necessary.

IX.

The Mamma.

Inflammatory diseases of the female breast are
frequent, painful, and depressing. In *acute inflammation,* and in *abscess,* a zone of iodine liniment

around the abscess should be combined with a stripe
either of iodine liniment or acetum lyttæ over the
brachial artery.   Over the whole of the inflamed part
a thick heavy poultice should be applied, with as much
pressure as is possible without producing pain.   The
pressure may be effected by a sort of many-tailed
bandage (made of three or four broad pieces of calico
or flannel), so arranged as to envelop the thorax
without moving the patient; or, if the patient is kept
in a completely horizontal posture, a shot mattress may
be placed over the poultice, and the outer side of the
breast supported by a cotton wool cushion in the
axilla.   Pressure, carefully exerted, is of the greatest
advantage in these cases.   Much of what was said on
the treatment of abscess applies to mammary abscess
(in whatever part of the organ it may be situated),
especially as regards the use of the knife.   As counter-
irritation relieves pain and subdues inflammation, there
need be at least no hurry in making an incision.

In the more *chronic abscesses*, an iodine pigment may
be applied freely around the inflammation and along the
inner side of the arm, combined with cotton wool or
poultice-pressure.   In chronic inflammatory thickening,
a very successful method of treatment is to strap the
breast, and apply iodine over the brachial region.
Hyperthrophy of the breast, when affecting the whole
gland, is much more effectively treated by counter-

irritation to the inside of the arm, in addition to
pressure, than by strapping alone. Similar treatment
is also the best for the clusters of enlarged glands
that are occasionally met with between the mamma
and the axilla.

The Vulva.—In *vulvitis*, whether simple or follicular,
or gangrenous, in infantile leucorrhœa, and in abscess,
the production of counter-irritation, in the form of a
patch in each groin, or a stripe over the femoral
artery, will more quickly than any other remedy cut
short the disease. Acetum lyttæ, or iodine liniment,
or pigment, or tincture may be required, according to
the severity of the case, or the age of the patient; if
the acetum lyttæ be used, or the undiluted iodine
liniment, I prefer the stripe over the femoral.

The Vagina.—In *vaginitis*, and in *gonorrhœa* in the
female, the treatment is similar to that just described
for inflammatory conditions of the vulva, and to that
of gonorrhœa in the male. In acute cases, the acetum
lyttæ, in a stripe along the femoral artery, is very
efficient. The iodine preparations may be used, but
over a larger surface. When the stripe of acetum
lyttæ is used, a disc of iodine pigment may be added,
the centre of the disc being at the vulva. This
treatment is very rapid in its results, but, should any
delay occur, astringent injections are so readily
applied, that their use should be resorted to.

XCIII.

Sub-acute prostatitis : Iodine liniment to perinæum, thighs, and groins : Immediate relief.—George T., aged 34, married, had gonorrhœa ten weeks before. Frequent micturiton, with spasm of neck of bladder, weight, and pain, and tenderness, in perinæum, cloudy urine, and a gleety discharge came on. Iodine liniment was applied in a disc around the genital organs. In three days the pain, tenderness, and vesical irritability were completely removed.

Case XCIV.

Chronic prostatitis : Counter-irritation : Rapid relief. —A gentleman came to me with perinæal pain, with vesical irritability, and tenesmus, succeeding a gonorrhœa. Iodine to the perinæum and inner sides of thighs, removed the symptoms in forty-eight hours. He had been undergoing for some time the ordinary medical treatment.

Case XCV.

Acute prostatitis and cystitis (of neck of bladder) : Circle of iodine around the genitals : Immediate recovery. —Samuel P., aged 42 ; during an attack of gonorrhœa the micturition became very frequent, and the stream suddenly slower. There was acute perinæal pain, before and after micturition. The urine was cloudy. A circle of iodine liniment was applied around the genitals, with the most marked relief in twenty-four hours.

Case XCVI.

Gleet of two months: Counter-irritation: Recovery in one week.—John Q., aged 40, single, had gonorrhœa and gleet two months. The iodine liniment was applied over the femorals. An injection of chloride of zinc was also used. Recovery was complete in a week.

Case XCVII.

Severe chordee: Counter-irritation successful, when other modes of treatment failed.—Mr. W., aged 22, had severe and painful chordee, which was quite unaffected by the ordinary internal and local treatment. The repeated application of narrow lines of acetum lyttæ at the under surface of the penis, a narrow stripe being applied every three or four days in a new locality. The relief was immediate and progressive, the pain ceased in a few nights, but some weeks elapsed before the curvature quite disappeared.

Case XCVIII.

Severe gonorrhœa cured in two days by counter-irritation.—Mr. B., aged 19, had gonorrhœa for the first time, and suffered extremely from scalding and discharge. Two stripes of blister over the femorals removed the scalding completely in twenty-four hours, and the discharge in forty-eight. *Remarks.*—Blisters I now use only in exceptional cases. Acetum lyttæ or iodine liniment over a large surface is better.

CASE XCIX.

Chordee after Holt's operation, much relieved by successive small patches of acetum lyttæ in the perinæum.—Mr. P., 32, single, after Holt's operation, which answered its purpose very successfully, was for two or three nights constantly awoke with painful erections. Camphor, opium, and belladonna were useless. A little patch of acetum lyttæ in the perinæum each night gave immediate and great relief.

CASE C.

Gonorrhœa treated with blisters: Well in six days.— James B., aged 23, had gonorrhœa eight days, severe attack. Two small blisters were applied over femorals. Scalding disappeared in one day, and discharge completely ceased in five days.

CASE CI.

Gonorrhœa with numerous warts: Cured rapidly by counter-irritation.—Henry H., aged 19, had gonorrhœa

several weeks, and recently a large crop of small warts
had appeared around the corona. Iodine paint was
used in a circle to the pubes, thighs, and perinæum.
The gonorrhœa was quickly well, and the warts
gradually disappeared in about three weeks.

Case CII.

Gonorrhœa : Iodine irritation : Well in a week.—
Thomas W., aged 19, came with gonorrhœa, of ten
weeks' duration. First time. With a circle of iodine,
maintained with vigour, was well in one week.

Case CIII.

Gonorrhœa : Iodine irritation : Well in a week.—
John N., aged 19, had gonorrhœa one month. Much
scalding and discharge. With an iodine circle, was
well in one week. *Remarks.*—Severer counter-irritation (acetum lyttæ, say) is not desirable where the
inflammation has become chronic.

Case CIV.

*Gonorrhœa and consecutive orchitis: Counter-irritation:
Recovery in three days.*—Henry A., aged 17, had
gonorrhœa ten days and orchitis one day. Two broad
stripes of iodine liniment over femorals, and nitrate of
silver over testis, removed the orchitis in twenty-four
hours, the gonorrhœa in three days.

Case CV.

Gonorrhœa (severe): Counter-irritation: Scalding removed immediately, and discharge in a week.—David E., aged 58, had gonorrhœa three weeks, the scalding and discharge having no tendency to diminish. A circle of iodine liniment removed the scalding immediately, and, the irritation being maintained, the discharge entirely ceased in a week.

Case CVI.

Chordee, with much induration of spongy body: Counter-irritation : Recovery.—John M., aged 21, had gonorrhœa and extensive induration of spongy body, especially near glans. The application of a circle of iodine around the genitals and a band around the penis quickly removed the gonorrhœa, and then the induration.

Case CVII.

Gonorrhœa in the female: Counter-irritation: Recovery in four days.—Sarah S., aged 22, single, had had gonorrhœa thirteen weeks. Some scalding and very copious discharge. Iodine paint was applied in a circular patch in the groin, and prolonged over the femorals. In four days, she was completely well.

v

Case CVIII.

Gleet : Counter-irritation : Rapid recovery.—George R., aged 17, suffered from gleet of many months' duration. A circle of iodine removed the discharge in three days. A little scalding, curiously enough, remained a few days longer.

Case CIX.

Gonorrhœa, of six weeks' duration : Counter-irritation : Rapid recovery.—Alfred E., aged 22, had gonorrhœa six weeks. Had two attacks before. Discharge too copious, and scalding too severe to permit the use of the bougie. A circle of iodine-irritation, of a mild character, was ordered. In fourteen days, the recovery was complete. *Remarks.*—Where there is a slight stricture, the counter-irritation should be mild; but, even in these cases, until the bougie can be used, it is the best remedy.

Case CX.

Gonorrhœa (second) : Counter-irritation : Well in six days.—James B., aged 21, had gonorrhœa of three weeks' duration. An attack twelve months before. A circle of iodine was followed by recovery in six days.

Case CXI.

Gonorrhœa, implicating the uterus : Striking result of counter-irritation.—A woman came to the hospital with old-standing gonorrhœa. Her appearance was pale and haggard. Sacral and abdominal pains were frequent. A narrow stripe of blister was put over each femoral. She was to rest as much as possible. In six days, she came again, and it was difficult to believe she was the same woman. She was rosy, stout, well, and looked ten years younger.

Case CXII.

Gonorrhœa : Counter-irritation : Recovery in six days.—William L., aged 28, had gonorrhœa before, which lasted twelve months, with copaiba and ordinary treatment. Second attack ; duration, three weeks. Circle of iodine, and alkaline and iron mixture. Well in six days.

Case CXIII.

Gonorrhœa: Iodine counter-irritation: Rapid recovery. —Edwin R., aged 20, had gonorrhœa one month. Discharge very copious. A disc of iodine paint, mildly maintained, removed the symptoms in five days.

Case CXIV.

Gonorrhœa: counter-irritation: Recovery in four days.—Henry J., aged 22, had gonorrhœa of five weeks' duration. A disc of iodine irritation removed the symptoms in three days.

Case CXV.

Gonorrhœa: Counter-irritation: Recovery in three days.—Thomas D., aged 32, had gonorrhœa three weeks. With iodine irritation was well in three days.

Case CXVI.

Gonorrhœa of six months' duration: Counter-irritation: Recovery in ten days.—Amos S., aged 18, came with gonorrhœa of six months' duration, and probably slight stricture. Scalding and discharge marked. Counter-irritation over the femorals was followed by recovery in ten days.

Case CXVII.

Gleet and stricture: Counter-irritation: Gleet removed in seven days.—William D., aged 36, single, came with gleet and stricture. Had had gonorrhœa several times before. Counter-irritation to perinæum and femorals completely removed the discharge in six days.

Case CXVIII.

Gonorrhœa (probably first): Counter-irritation: Recovery in three days.—Walter H., aged 20, came with copious discharge and sharp scalding, of twelve days' duration. Perinæal and femoral counter-irritation was followed by recovery in three days. *Remarks.* —The rapidity of the recovery, where there is no stricture, is well seen here.

Case CIX.

Gonorrhœa, in a married woman, cured in three days with counter-irritation.—Mrs. L., aged 28, came with gonorrhœa of long standing, unrelieved by every kind of injection. Two narrow blisters over the femorals were followed by complete recovery in three days.

Case CXX.

First gonorrhœa cured in three days by small application of iodine.—Joseph B., aged 18, came with gonorrhœa of one week's duration. Iodine liniment to penis, perinæum and femorals applied once, freely, removed all symptoms in three days.

Case CXXI.

Double acute orchitis (very severe): Counter-irritation to scrotum and femorals: Recovery in twenty-four hours.

—A man, with severe double orchitis, unable to stand upright, and in great agony, applied for admission into the hospital   The scrotum was covered with a solution of arg. nit., and iodine liniment was applied over the femorals.   Recovery was complete in twenty-four hours.

Case CXXII.

Acute orchitis: Counter-irritation: Recovery in less than forty-eight hours.—Mark D , applied with gonorrhœa, of three weeks' standing, and acute orchitis.   Arg. nit. was applied to the scrotum of the affected side, and iodine liniment over the femoral. All the symptoms passed away in less than forty-eight hours.

Case CXXIII.

Acute orchitis: Counter irritation: Rapid recovery.— George L., aged 27, came to hospital with acute orchitis.   Arg. nit. solution, not very strong, was applied.   In two days, all discomfort had subsided, but a little enlargement remained, which was removed in a few days by iodine to the femorals.

Case CXXIV.

Acute orchitis: Counter irritation: Recovery in twenty-four hours.—Charles C., aged 26, applied with severe consecutive orchitis.   Arg. nit, thoroughly to scrotum, and iodine to femorals, removed all symptoms in twenty-four hours.

CASE CXXV.

Acute orchitis : Counter irritation : Recovery in forty-eight hours.—Alfred C., aged 22, applied with acute orchitis, not of the severest kind. A milder counter-irritation relieved the symptoms in forty-eight hours.

CASE CXXVI.

Gonorrhœa and acute orchitis :   Counter-irritation : Immediate recovery from both.—Edward F., aged 19, applied. Acute orchitis and gonorrhœa, with severe scalding and discharge.   Counter-irritation in the usual way instantaneously relieved the scalding.   In less than forty-eight hours, the orchitis had completely, and the discharge almost, disappeared.

CASE CXXVII.

Acute orchitis: Counter-irritation over femoral arteries: Recovery in forty-eight hours.—John B., aged 25, applied with acute orchitis, following a gleet. A single free application of iodine liniment, over the femorals, removed all the symptoms in forty-eight hours.

Case CXXVIII.

Acute orchitis: Counter-irritation: Recovery in twenty-four hours.—Francis C., aged 30, applied with severe orchitis. Nitrate of silver to scrotum, and acetum lyttæ over the femorals immediately relieved the pain, and removed all the symptoms in twenty-four hours.

Case CXXIX.

Discharging abscess of breast of two months' duration: Counter-irritation: Recovery in ten days.—Phoebe E., aged 33, applied to the hospital with a painful, freely suppurating abscess of the left breast. Gentle pressure was ordered for the breast itself, and iodine liniment was applied to the inner side of the arm. All the symptoms disappeared in ten days.

Case CXXX.

Mammary abscess: Opening and closure much accelerated by counter-irritation.—Sarah F., aged 25, applied with an unopened abscess of the breast, of six weeks' duration. There was much induration and pain, with emaciation and sleeplessness. Iodine liniment was applied over the brachial artery, and a linseed poultice, with a bandage, to the breast. Pain ceased immediately, and prolonged sleep was obtained.

In a few days the abscess opened spontaneously, and six days later it was well.

CASE CXXXI.

General hypertrophy of breast : Recovery from counter-irritation to arm.—A woman, aged 22, single, applied with general hypertrophy of left breast, which was three times the size of the right. Strapping and internal remedies, after a long trial, failed to do any good. Counter-irritation was then effected over the brachial artery, with marked and rapid benefit.

CASE CXXXII.

Abscess of breast of five weeks' duration : Counter-irritation over brachial : Recovery by absorption in a week.—Ellen E., aged 20, had a tender, painful, red swelling of the breast, of five weeks' duration. Strapping to the breast and iodine liniment over the brachial artery removed all the symptoms in a week.

CASE CXXXIII.

(Under Dr. Hodge's care).—*Discharging abscess of the breast : Adjacent counter-irritation : Recovery in three days.* — "On January 31st, M. D., aged 26, married (four children), was brought to me by her mother, one of our nurses, suffering from abscess of right breast. The mamma was twice its natural size,

hard, and infiltrated. There was an opening about the size of a sixpence, from which very unhealthy looking pus was exuding. It had existed nine days, and had been repeatedly poulticed without relief. Patient stated that her sufferings had been so great, that she had scarcely had an hour's sleep for the past week, and her looks fully confirmed her statement, for a countenance more expressive of pain and continued sleepless nights, I never saw. Taking the margin of hardness as the point beyond which I should not go, I made a semi-circular ring on the upper part of the affected breast with the liniment of iodine. I continued the iodine well over the corresponding clavicle, and extended the paint laterally, and thus included a pretty extensive surface. I enveloped the breast in cotton wool, and then applied, as firmly as possible, without causing pain, a flannel bandage, with a view not only of taking off the weight of the breast, but of exciting compression upon it.

"On Feb. 3, my patient called to see me. I was struck with her altered appearance—the pale, haggard look was gone, and she had quite a colour in her cheeks. Upon my asking her how she was, she said, with a smile, "O, sir, I think I am well." Scarcely crediting quite so satisfactory a condition, I proceeded to examine the breast, and found her statement literally true, the hardness and infiltration was away, the right being as soft and resilient to the touch as its fellow ; both breasts were identical in size. The cicatrization was the only means by which we could then have told which breast had been affected with abscess.

" Patient informed me that she had not had the slightest pain or uneasiness since I had applied the iodine, and had slept well every night. I may add, that her gratitude was beyond description."

Remarks.—I have given this case in Dr. Hodge's own words. The records of surgery contain nothing more miraculous.

Cases under Mr. Turton's care.—Mr. Turton remarks as follows :—" In several cases of acute inflammation of the mammary gland, which would undoubtedly have ended in suppuration, a free application of a strong solution of iodine around the breast has invariably relieved the pain and other symptoms of inflammation. I think that if it is applied early, the result will always be to prevent suppuration In half-a-dozen cases only one gathered. This I opened, and I certainly never before saw one of these cases get well with so little discharge or so little pain."

XX.

Inflammatory Diseases of the Eye.

On the subject of diseases of the eye I shall be very brief. It is a department of surgery more advanced than any other, and quite recently several admirable text books have brought a thoroughly modern knowledge of it within the reach of the busiest practitioner. Many of the diseases, especially the deeper ones, are of a specific character, and require specific treatment. Very many are of a degenerative rather than inflammatory type. Many are successfully treated only by operative interference.

There are a few diseases, however, where inflammation, as such, is destructive to sight, as gonorrhœal ophthalmia, purulent ophthalmia, purulent ophthalmia of infants, and the deep and sloughing ulcers of the cornea. In these cases, a ring of counter-irritation around and at some distance from the eye, is usually rapid—often magical in its effects. In *gonorrhœal ophthalmia*, at any stage so long as the cornea is still living, a circle (interrupted or not at the nose) of acetum lyttæ, three-quarters of an inch or an inch wide, will save the cornea. If it be necessary, the effect can be kept up by a second, or even a third, circle, within or without the first. In the *purulent ophthalmia* of infants, a similarly interrupted circle may be made with nitrate of silver or iodine, or, if it be a narrow one, with acetum lyttæ. The purulent ophthalmia of adults should be treated as gonorrhœal ophthalmia, with such modification as the mildness or severity of the disease may require. Phlyctenular ophthalmia is at the present time very commonly treated by counter-irritation. I have repeatedly seen eyes, which have been for weeks or months affected with urgent photophobia, comfortably open a few hours after the application of nitrate of silver around the orbit. In severe and relapsing cases, a circle of acetum lyttæ, well removed from the eye, is a remarkably successful treatment. Ulcers of the cornea, especially the deep and the sloughing, are also best treated by circumorbital counter-irritation.

In other diseases of the eye, where the inflammatory process is a dominant feature, it is well that the surgeon should bear in mind the potency of counter-irritation. It has the merit of not interfering with other measures, such as the use of atropine and other local remedies, or the internal administration of mercury or other drugs.

Diseases of the ear are, in a very large proportion, of an inflammatory character. Where the inflammation is recent, whatever the degree of acuteness, and whether situated in the meatus, or the tympanal membrane, or in the tympanal cavity, the most satisfactory results are obtained by counter-irritation. Practically, we have no other remedy in these cases, whether we use the cantharides, or iodine, or mustard, or hot water and linseed meal. In acute cases I place, with marked benefit, a horseshoe under and around the ear. In chronic inflammations (and these form the great majority of diseases of the ear), the counter-irritation may be applied to the mastoid process and carried to the back of the neck. An incision over the mastoid process, in acute inflammation of the membrana tympani, or tympanal lining, has been recommended; it acts of course as a counter-irritant, and a sufficiently large blister would act as well. In inflammatory thickening of the mucous lining of the eustachian tube, the application of iodine and nitrate of silver to the tonsil acts as counter-irritation to the eustachian tube.

THE END.

INDEX.

Printed by Corns and Bartleet, Union Street, Birmingham,

www.ingramcontent.com/pod-product-compliance
Lightning Source LLC
Chambersburg PA
CBHW021659210326

41599CB00013B/1469